An introduction to HF Software Defined Radio

By Andrew Barron ZL3DW

Edition 1.0 July 2014.

ISBN-13: 978-1500119935

ISBN-10: 1500119938

The information contained is my own analysis of Software Defined Radio as used by Radio Amateurs worldwide, it intends no endorsement of any particular brand of radio equipment and the author has no association with any equipment manufacturer or SDR software developer. Research material for the creation of this document has been sourced from variety of public domain Internet sites. The author accepts no responsibility for the accuracy of any information presented herein. It is to the best of my knowledge accurate but no guarantee is given or implied. Use the information contained in this document at your own risk. Errors and Omissions Excepted.

The OpenHPSDR group. *"The HPSDR is an open source (GNU type) hardware and software project intended as a "next generation" Software Defined Radio (SDR) for use by Radio Amateurs ("hams") and Short Wave Listeners (SWLs). It is being designed and developed by a group of SDR enthusiasts with representation from interested experimenters worldwide,"* see (http://openhpsdr.org/).

An introduction to HF Software Defined Radio

By Andrew Barron ZL3DW

Table of Contents

This page was intentionally left blank.

Chapter 1 - Introduction to Software Defined Radio

One of the more recent developments in amateur radio has been the proliferation of ham radio Software Defined Radio (SDR) receivers and transceivers. SDR is a logical development of the DSP (digital signal processing) technology already found in many ham radio transceivers. It has attracted a large number of devotees and probably an equal number of critics. I say, "Don't knock it 'til you've tried it."

But, what is it? How does it work? Is it really better? Do you need to be a computer or technology boffin? Should you buy one or should you stick to a standard radio? There are a lot of questions to answer and a lot of the information on the Internet is out of date or just plain uninformed, so I decided to write a book dealing with different aspects of SDR technology.

At the moment SDR is not for everyone but it is going main stream and there is a good chance that the next transceiver you buy will be an SDR. Many hams think SDRs are a printed circuit board kit or a black box radio with no knobs. Take a look at the photo of the Elecraft KX3. It is an SDR, it has lots of buttons and a few knobs and you don't need a PC to use it. The KX3 was designed for QRP and portable operation with a maximum output power of 12 Watts. It has one of the best HF receivers available at any price; coming in near the top of the Sherwood Engineering list and 5th based on my averages of the ARRL lab tests. Baofeng makes a VHF/UHF dual band handheld which is an SDR. Elecraft, TenTec, Alinco and ADT all make HF ham band SDR radios 'with knobs'. It is becoming cheaper to make medium to high performance radios using SDR technology than to make traditional radios. So I am sure we will see more SDR based transceivers released in the next few years.

The Elecraft KX3 transceiver is a Software Defined Radio, 'with knobs'. It can be used without a connection to PC. An optional PC connection provides features such as CAT control, digital mode operation and a band scope display.

So now there are two types of SDR, those which have knobs and don't need a PC, and those which don't have knobs and do need a PC. The underlying technology is the same in both and it is radically different to conventional ham radios.

I never thought I would ever graduate to a radio with no knobs. I believed, "The more knobs the better," but I started experimenting with SDR because I wanted one of those fancy band scopes which were appearing on the latest high end ham transceivers. I purchased a single band Softrock kit on 9 MHz and connected it to the 9 MHz IF on my old FT-301. I was not about to mess with my newer Yaesu radio, besides its 45 MHz IF was too high for a Softrock kit. It worked very well, I was amazed to be able to see SSB and CW signals as well as hear them.

There was a significant delay or 'latency' in the audio from the Softrock compared to the audio direct from the Yaesu radio due to the slow processing in my old Pentium 4 PC. But as long as you listened to either the radio or the SDR it was fine. The results were so impressive that I moved on to a Softrock transceiver kit then I bought a FLEX-1500 QRP SDR radio. At the time I still thought the Yaesu would be my main radio and the Flex radio just a toy to experiment with. Since then I have become hooked on the SDR 'panadapter' band scope display, the remarkable performance and the ease of using the SDR transceiver. Now when I operate a conventional radio I feel deprived. It feels like I am operating blind.

Software defined radio has been around for a while especially in cellular radio base stations and military radio applications. In the ham radio world, the Tayloe detector was patented in 2001, the high performance SDR (HPSDR) group was formed in 2005 and the first commercial ham SDR transceiver, the FlexRadio SDR-1000 was released in 2006. It is still fairly new, but hams have always been ready to adopt new technologies and have frequently been involved in developing them. The first amateur radio satellite was launched only four years after the first ever man made satellite. Amateur TV, digital modes, packet, satellite, LF, EME, Echolink, IRLP, APRS, microwave, meteor scatter, repeater operation and even SSB are all examples of what were once new ways to experiment with ham radio.

I guess SDR is a product of the 'digital age'. We kept our LP records and some hams refuse to use that "new fangled" SSB mode, but most of us now own a personal computer, use a digital cell phone and listen to CDs. Some people still think 'vinyl' is better, the world is flat and SDRs are no good. I can't change that, so this book is for everybody else.

When DSP (digital signal processing) was first introduced on ham transceivers it worked on the audio signal after the final mixer. The audio is converted to a digital signal, processed in the DSP chip, then converted back to an analog signal and sent to the audio amplifier and speaker. This method is now referred to as AF DSP (audio frequency digital signal processing). It provides 'high cut', 'low cut', noise filters and often a notch filter. The main problem with AF DSP is although you can notch out the guy tuning up 3 kHz up the band, or set the filters so you can only hear the one CW signal you want to work, the unwanted signals are still within the IF pass band and they affect the AGC action of the receiver. This is a major problem when I work PSK stations with my Yaesu radio which only has AF DSP. Every time a high power station transmits, the weak stations I want to work fade away.

To eliminate this problem, the next generation of DSP worked on a low IF frequency. Again the signal is converted to a digital signal, processed in a DSP chip then converted back to an analog signal. IF DSP adds dynamic 'IF width' and 'IF Shift', or a range of pre-set IF filters and often a band-scope, to the AF DSP features. Between the IF and the audio section lies the demodulator, or in the transmit chain the modulator. The logical next step is to carry out those functions using digital processing as well.

Definitions vary, but in general terms, if the filtering and demodulation (or modulation) is done using software which is re-configurable without replacing chips in the radio, then the radio is 'Software Defined'. In other words if the signal is converted to a digital signal and processed using reconfigurable software or firmware running on a PC, embedded computer processor, or FPGA (field programmable gate array) it is a software defined radio. But if the signal is converted to a digital signal and processed in a dedicated (not reconfigurable in the radio) DSP chip it is not a software defined radio.

The old definition is not really adequate anymore because the lines are blurred. Many conventional radios can have firmware updates making them at least partially field upgradable and radios like the Elecraft KX3 are considered to be SDRs even though they use dedicated DSP chips. These days what is, and what is not, an SDR is more usually defined by the hardware architecture. Conventional radios use a double or triple conversion superhetrodyne design with front end filters, hardware oscillators, mixers, and IF amplifiers, usually followed by dedicated DSP chips. SDRs use Direct Digital Sampling or Quadrature Sampling Detectors and Exciters. I cover how these work in the following chapters.

SDR in the ham world is evolving rapidly and it is splitting into several streams. There are SDRs being developed that look and work like any other ham radio with all of the usual knobs and buttons. They are the, "same on the outside but different on the inside". They usually don't need a PC for their operation but in most cases they can be connected to one; for software updates, CAT control, band scopes, and digital modes. These radios are being developed because it is cheaper to get high levels of performance and a good range of functions using SDR architecture than it is to build high performance conventional radios. Then there are 'black box' radios with no knobs which do need a PC to provide the operating interface. They are mostly SDRs although some WinRadio and Icom receivers are not. The black box SDRs which need a PC, range from simple Softrock kits, to USB dongle receivers, small box receivers and QRP transceivers, up to 100W transceivers. The latest SDR models use direct digital sampling and offer some completely new features such as the ability to monitor several bands at the same time. Both kinds of SDR use the same type of hardware, but the discussion on SDR software for the PC relates mostly to the black box radios. Most of the commercially available software defined radios are receivers because the real benefits of software defined radio are exceptional receiver performance and operating features. Also, receivers are easier to design and build than transceivers. Currently, there are some QRP transceivers available and only a few 100W transceivers.

The PC software is a key part of the experience of using a software defined radio. It offers a user interface quite different to a conventional radio and you have to get used to operating in a different way. It is really no different to learning how to operate any new receiver or transceiver. Most ham radio transceivers produced over the last 20 years or so can be controlled using PC software. PC control is handy for things like operating digital modes, logging, and CW keyers. If you use CAT (Icom CI-V) or digital mode software you should have no trouble adjusting to using a SDR. But the PC software for an SDR is much more than just an interface to the radio, it usually performs part, or all, of the digital signal processing, including the filtering and demodulation, and in transceivers the modulation. One of the big advantages of using an SDR is the ability to try different software with the same radio and most of it is free. New software completely changes the way the radio looks and operates, it can add new modes and features. It is like getting a whole new radio each time you try a new SDR application and it helps to keep your radio completely up to date. If someone invents a new feature like transmitter pre-distortion or a new noise filter, you can often add it at no cost.

First generation SDRs known as, 'sound card SDRs', are cheap because most of the radio is implemented in the PC software. In many cases the hardware part of the radio is little more than a QSD (quadrature sampling detector), often with a Si570 chip working as a local oscillator to give an extended frequency range.

The QSD detects the RF signal and produces I and Q audio streams. Then the sound card in the PC is used to perform the analog to digital conversion (sampling). The PC software also performs the digital signal processing and provides the user interface. Although the performance of this type of SDR is really quite good there are some issues. PC sound cards are not designed for software defined radio. They add noise and distortion when sampling low frequencies below about 100 Hz and the maximum available sample rate of the PC sound card limits the bandwidth of the displayed spectrum and waterfall. This type of SDR is a direct conversion receiver. RF signals at the local oscillator (clock) frequency are converted to a 0 Hz audio frequency and are displayed in the middle of the PC display window with signals below the receiver local oscillator frequency shown to the left and signals above the receiver local oscillator frequency shown to the right. Typically the spectrum display from a sound card SDR will have a large noise spike in the middle of the spectrum display. This does not affect the receive performance when you select signals at other parts of the display, but it is annoying if you are using the SDR as a band scope connected to a receiver IF, because the signal you are listening to is right where the noise spike is displayed. Sometimes an 8 kHz or 10 kHz IF is used to offset the noise spike problem.

Second generation SDR receivers have a hardware ADC (analog to digital converter) in the radio and second generation SDR transceivers have the DAC (digital to analog converter) inside the radio as well. Faster and better A/D (analog to digital) conversion allows a wider bandwidth to be displayed on the spectrum display and the 0 Hz noise spike is eliminated because the on board ADC chips have better performance at low audio frequencies than average PC sound cards. Using an A/D conversion with more digital bits representing each sample creates more individual voltage levels so a wider dynamic range can be represented without overload. This allows the receiver to cope with large signals without being insensitive to weak signals.

Doing the digital signal processing in the PC places an additional load on the computer CPU and the PC uses a multi-tasking operating system which is not ideal for handling a continuous data stream from the radio. Receive and transmit performance could be affected by the specification of the PC and the demands of other software running at the same time. Don't let this put you off, I don't have a particularly fast PC, it is an old dual core, and I often run digital mode software, SDR software, a logging program, an Internet web browser, antenna rotator control, and several USB devices concurrently with no problem at all. There is also the problem of different PC configurations causing support issues for the radio suppliers. To eliminate these problems and to extend the performance of the radios there is a trend in 4th generation SDRs towards moving all or some of the digital processing back into firmware inside the radios. This is usually done by adding an FPGA (field programmable gate array) and sometimes DSP chips to the radio board. Sections of the FPGA can be configured to perform many tasks in parallel. 4th generation SDRs use direct digital sampling rather than a quadrature detector and they almost always include at least one on board ADC and an FPGA. The FPGA in a 4th generation SDR may only be used for filtering and decimation, but its functions could also include demodulation, modulation and DSP functions.

3rd generation SDRs are a branch of the 2nd generation SDRs. They still use a QSD rather than using digital down conversion but they also include on board ADC and DSP functions. Many of the SDRs 'with knobs' are 3rd generation SDRs.

Chapter 2 - The QSD method

Designing and producing high performance amateur radio receivers and transceivers is an exceptionally difficult task. Every amplifier stage in a receiver adds noise to the signal and this degrades the noise figure. Non-linear devices like mixers cause intermodulation distortion and local oscillators can add noise and harmonic related signals. The radio manufacturers have built on many years of experience and work very hard to minimise these effects. There is fierce competition among the 'big three' to produce transceivers with excellent receiver performance. Receiver IF frequencies are chosen to reduce the image frequencies and place them outside the ham bands. New receivers have very stable local oscillators with low phase noise. Some new receivers have roofing filters, variable pre-selectors, or external microprocessor controlled Hi-Q front end filters to make the receiver selectivity track the wanted receive signal rather than just having a wide roofing filter.

A large part of the reason that SDR receivers have very good performance is that the SDR design eliminates the need for multiple mixers, local oscillators and IF amplifiers. Eliminating these receiver components removes the noise and distortion they cause. All of the 1st generation 'sound card' and 2nd generation (A/D conversion in the radio) SDR receivers use a QSD (quadrature sampling detector) which acts like a direct conversion mixer. In most cases they use a circuit known as a Tayloe Detector which was patented by Dan Tayloe N7VE in 2001. The same design can be used in reverse as a quadrature sampling exciter (QSE) to make a signal which can be amplified to create a transmit signal. Gerald Youngblood AC5OG, the CEO of FlexRadio stated in a March 2003 QEX article that the Tayloe detector is the same in concept as modulator designs published by D. H. van Graas, PAØDEN, in 1990 and by Phil Rice, VK3BKR in 1998. However he also says, *"Traditional commutating mixers do not have capacitors (or integrators) on their output. The capacitor converts the commutating switch from a mixer into a sampling detector (more accurately a track-and hold)".* Anyway Dan holds the patent and most if not all ham radio QSD based receivers use the Tayloe design.

In any discussion about software defined radio it is not long before I and Q streams are mentioned. It is at this point people's eyes glaze over and they decide that SDR is too complicated for them. Don't panic, it is not really too difficult. But before we get into how the QSD design works I should explain about direct conversion receivers and 'the phasing method'. QSD based SDRs are direct conversion receivers. This is not a new idea. Direct conversion 'phasing' receivers such as the Central Electronics CE-100v were very popular in the mid-1950s until improvements in filter design made superhetrodyne receivers the preferred design.

Direct conversion receivers are similar to superhetrodyne receivers except the IF output is directly at audio frequencies extending from 0 Hz up to the bandwidth of the receiver. The local oscillator is at the same frequency as the wanted receive frequency, so in the mixer an upper sideband signal extending to 3 kHz above the local oscillator (LO) frequency becomes an upper sideband audio signal at 0 to 3 kHz. This is great except it is a mixer so it works both above and below the LO frequency. An RF signal 1 kHz below the local oscillator frequency cannot become -1 kHz and is reflected into the audio range with a phase change of 180 degrees, interfering with the wanted signal. If it was an upper sideband RF signal it becomes a lower sideband audio signal due to the 180 degree reflection.

Traditionally this image frequency problem is managed in one of two ways; either the signals below (or above) the LO frequency are filtered out before the mixer, in the same way that image signals are filtered out before a mixer in a superhet receiver, or the 'phasing method' is used to eliminate the image frequencies. SDR receivers use the phasing method for image frequency cancellation.

In the phasing method the signal is split into two streams before the mixer and one of them is delayed by 90 degrees. They are called "I" incident and "Q" quadrature signals. After the mixer a 90 degree phase change is applied to the Q stream. When the two streams are combined again, the wanted signals are back in phase and add together. But the unwanted image signals have a 180 degree phase change so the new 90 degree phase shift causes the two streams to become out of phase and they cancel.

The Tayloe QSD detector or 'switching integrator' creates an I stream and a Q stream at audio frequencies. The Q stream is generated from samples of the input signal taken 90 degrees after the I stream samples. After the detector the two audio signals are converted to digital data using the PC sound card or a dedicated analog to digital converter (ADC) chip. Software in the PC creates the second 90 degree phase change, allowing image cancellation using mathematics rather than circuitry.

The Tayloe QSD detector is very simple and cheap to make. It uses three very basic integrated circuits, a dual Flip Flop latch configured to divide the clock signal by four, a multiplex switch chip and a dual low noise Op-Amp.

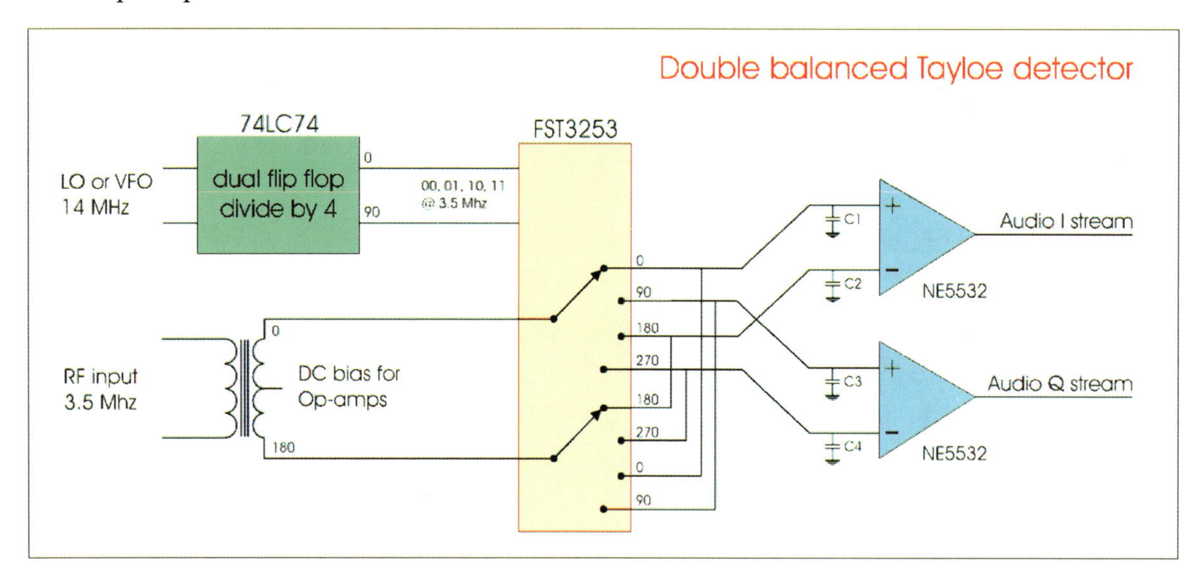

Take a look at the schematic diagram of the slightly more complex 'double balanced' Tayloe detector above. The two flip flops in a 74LC74 chip are configured to divide the 14 MHz clock signal by four. The two outputs generate a 00, 01, 10, 11 binary pattern which is used to switch the two multiplex switches to each of the four outputs in turn. If the clock VFO signal is from a variable device like a Si570 chip the receiver can be made to cover a wide range of frequencies. The Tayloe design can be used up to around 1 GHz, but in this example the clock signals to the switch are at 3.5 MHz and the received signals are also centred around 3.5 MHz. This means that the signal from the input transformer is switched to all four outputs during every cycle of the input frequency. A pulse of the input signal voltage is applied to each of the four capacitors in turn.

The capacitors store the voltage on the inputs to the Op-amps and end up averaging the input levels over time. It works as an envelope detector in the same way that the capacitor following the diode in a crystal set does. In the double balanced Tayloe detector, the signal from the other side of the input transformer is not wasted. Because it is anti-phase it is used to top up the capacitor that is 180 degrees out of phase. So for each of the four switch positions two of the capacitors are charged up with the input signal. As Dan Tayloe states in his article titled "Ultra Low Noise, High Performance, Zero IF Quadrature Product Detector and Preamplifier", *"... two separate detectors are driven with the inputs 180 degrees apart using an input transformer. The two detectors use a common set of four detector caps. Since the outputs of the two detectors are 180 degrees out of phase, the detector capacitors are now driven two at a time"*. The bias voltage for the Op-Amps is applied to the transformer centre tap and passes through the switches to the Op-Amp inputs.

Because the signal on the 180 degree capacitor is essentially the same as the 0 degree capacitor but with reverse polarity the 0 and 180 degree signals can be combined in the Op-amp. This gives a 6 dB increase in the, I stream signal level without adding any noise to the signal. Likewise the 90 degree and 270 degrees are combined in the other Op-Amp to create the Q signal, also with a 6 dB enhancement. Just like the 1950s phasing receivers, both of the audio outputs contain the same signals but the Q signal is delayed by 90 degrees. The audio output signals extend from 0 Hz up to the bandwidth of the receiver.

The bandwidth of a Tayloe detector is limited by the RC time constant of the capacitors and the overall resistance of the network. Because in the double balanced version the capacitors get topped up twice as often they can be smaller values, so the bandwidth is wider than the original single balanced version. The other factor limiting the bandwidth is the sample rate of the analog to digital converter following the detector. In a 1st generation SDR the ADC is the PC sound card. The I and Q signals are connected to the Left and Right line input on the PC. Don't use the microphone input because it is often not stereo.

Harry Nyquist (1889-1976) determined the fundamental rules for analog to digital conversion. He found that as long as the sample rate was at least twice the rate of the highest frequency in the analog signal, the original analog signal could be recreated accurately from the digital data. Most PC sound cards can do a stereo analog to digital conversion at 48 ksps (48,000 samples per second), or 96 ksps (96,000 samples per second). Some elite sound cards can manage 192 ksps. To keep Mr Nyquist happy, this means the maximum bandwidth that can be sampled is 24 kHz, 48 kHz or maybe 96 kHz. But because the Tayloe detector is acting as a mixer, both audio streams contain signals from above the LO frequency and signals from below the LO frequency. Once these are separated our SDR receiver can display a spectrum 48 kHz, 96 kHz or 192 kHz wide. If a dedicated ADC chip is used, you can achieve a bandwidth greater than 192 kHz. In the PC software, the I and Q digital signals can be used to determine the amplitude and phase of the signals at the times they were sampled. The phase information is used to decode FM signals and the amplitude information on both streams is used to decode amplitude modulated signals like CW, AM and SSB. In fact if you have the I and Q signals you can demodulate any type of modulation.

A 48 kHz or 96 kHz bandwidth is enough to display quite a few SSB signals or the entire CW or digital mode section of the band. You can click your computer mouse on any displayed signal and hear the QSO. Another function unique to SDR receivers is you can record the full bandwidth and play it back later. You can listen to any of the QSOs on the recorded band segment even if you didn't earlier.

For signals above the LO frequency, the audio on the I stream leads the audio on the Q stream by 90 degrees, (remember that the Q stream samples were created 90 degrees later). Signals from below the LO frequency end up with the audio on the I stream lagging behind the audio on the Q stream by 90 degrees, I will explain why in the next section. This allows us to extract the signals from above the LO frequency without images from below the LO frequency and use FFT (fast Fourier transformation) to display them on the right side of the spectrum display and we can also extract the signals from below the LO frequency and display them on the left side of the spectrum display. In this way the full spectrum can be shown both above and below the local oscillator or 'centre frequency'.

To get good image cancellation the audio level of the I and Q streams must be the same and the phase difference must be exactly 90 degrees. 40 dB of image cancellation requires the levels to be within 0.1 dB and the phase to be within 1 degree. 60 dB of image cancellation requires them to be within 0.01 dB and 0.1 degrees. The PC software is able to compensate for phase and amplitude errors to minimise the display and reception of image signals but it is not perfect across the whole spectrum.

If you have a QSD type receiver you can see the effect of the image cancellation by disconnecting or turning down the level of either the I or the Q stream, thus disrupting the IQ balance. You will see for every signal there is a mirror image on the other side of the spectrum display equidistant from the centre frequency. The mirror images have the sideband reversed which is not a problem and not visible for CW, AM or PSK signals but is obvious with SSB signals.

The number of bits the ADC uses to code the data affects the dynamic range of RF signals the receiver can handle. The theoretical maximum for a 16 bit A/D conversion is 96 dB. For a 24 bit A/D conversion it improves to 144 dB. In the real world you can expect a maximum dynamic range of around 130 dB. Even the 16 bit performance is much better than the 75 – 80 dB dynamic range achieved by a typical conventional receiver, so the SDR does not need an AGC loop to limit strong signals. However most SDR software does include AGC, simply to limit the audio level on strong signals.

Some other statistics for the Tayloe detector include a conversion loss of 0.9 dB (a typical conventional receiver mixer has 6 to 8 dB), a low Noise Figure around 3.9 dB, (the NF of a typical HF receiver can be anything up to 20 dB), and a 3rd order intercept point of +30 dBm. It also has 6 dB of gain without adding noise and it is very cheap.

The QSD SDR provides very good receiver performance at a lower cost than conventional receivers and it achieves the goal of making the modulator and demodulator a part of the digital signal processing, bridging the gap between IF DSP and AF DSP. But it still has one mixing process which is a potential source of intermodulation distortion. The next logical step is to eliminate the QSD mixer and sample the RF spectrum directly at the antenna. This is the basis of 4th generation SDRs. The process is known as Direct Digital Sampling (DDS). It uses Digital Down Conversion (DDC) and for transmitters Digital Up Conversion (DUC).

DDC software defined radio receivers sample the entire HF spectrum from a few kHz up to 55 MHz or higher at once. To comply with the Nyquist theorem this requires very fast analog to digital converters which have only become available in the last few years.

Image cancellation

The Tayloe detector is a 'direct conversion' mixer meaning the RF signal at the input is converted directly to an audio output signal.

The problem with direct conversion is that 'image' signals from below the LO (local oscillator) frequency end up in the audio signal as well as the signals from above the LO frequency. We want to be able to see and hear signals from above and below the LO frequency without them appearing on top of each other. Luckily the I and Q data allows us to separate the image signals from the wanted signals.

For input signals above the LO frequency, the audio on the I stream detector output leads the audio on the Q stream by 90 degrees. This is because the Q stream samples were created 90 degrees later. In the PC software a +90 degree phase shift is applied to the Q stream. Which realigns the I and Q streams so they become in-phase. The two data streams are then **added** together to make a new audio stream with twice the original amplitude. See the vector diagram on page 11 for a picture of how the signals combine. Signals that are below the LO frequency are reflected back into the same audio range as the signals above the LO, but the reflection causes a 180 degree phase change. So, for input signals below the LO frequency the audio output on the I stream lags the audio on the Q stream by 90 degrees. When we apply the additional +90 degree phase shift to the Q stream, the I and Q signals become anti-phase and when the two streams are added together they cancel out. Provided the two audio streams are at exactly the same level and the phase change between the I and Q streams is exactly 90 degrees, we can completely eliminate these image signals. Now the cool part! If we **subtract** the I stream from the Q stream instead of adding them together, we end up extracting the signals from below the LO frequency and cancelling the signals from above the LO frequency. By using both the addition and subtraction methods we end up with the wanted signals from above and below the LO frequency with all of the image signals cancelled.

One of the problems with QSD receivers is that for perfect image cancellation you have to have the I and Q streams at exactly 90 degrees and at exactly the same levels. Unfortunately this is not possible using a hardware mixer like the Tayloe detector. Variances in the Tayloe detector components mean the IQ balance is not perfect across the frequency range of the device and it may also vary with factors like power supply voltage or temperature. Also; the PC sound card may not have a perfect level and phase response between the left and right audio channels.

To make the image cancellation as good as possible the PC software looks at signals you are receiving, works out where the image frequency would be and adjusts the phase between streams and the audio levels to minimise the image signal. It does this all the time the receiver is operating and over time it learns the characteristic of your particular receiver. In most amateur radio SDR applications the software code used for this adaptive correction is based on a very clever and remarkably simple algorithm invented by Alex Shovkoplyas, VE3NEA, the author of the *Rocky* SDR software. He writes; *"The algorithm works as follows – the receiver pass spectrum is scanned for signals that are at least 30 dB above the noise. For each signal, synchronous detection of the image is performed, using the main signal as a reference oscillator. The synchronous detector has very high sensitivity and can detect the image signal, even if is below the noise."*

This technique allows QSD SDR receivers to achieve around 90 dB of image suppression which is comparable to a very good Superhet receiver. Apparently SDR guru Phil Harmon VK6APH reverently referred to these lines as, "The four lines of code that changed the SDR world".

Direct digital conversion receivers use a mixer and local oscillator created in software running on the FPGA chip to create the I and Q audio streams. This creates a perfect phase and level relationship, so the image cancellation is excellent, at least as good as the 100 dB or so achievable in the best conventional architecture receivers.

Block diagrams of the SDR receiver

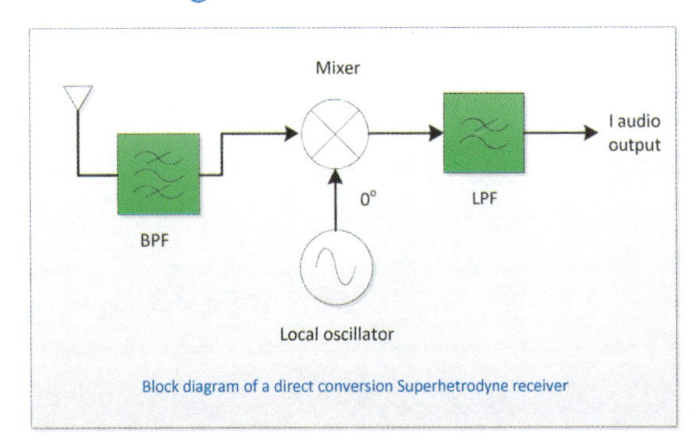

Block diagram of a direct conversion Superhetrodyne receiver

Above is the block diagram for a single conversion Superhetrodyne receiver.

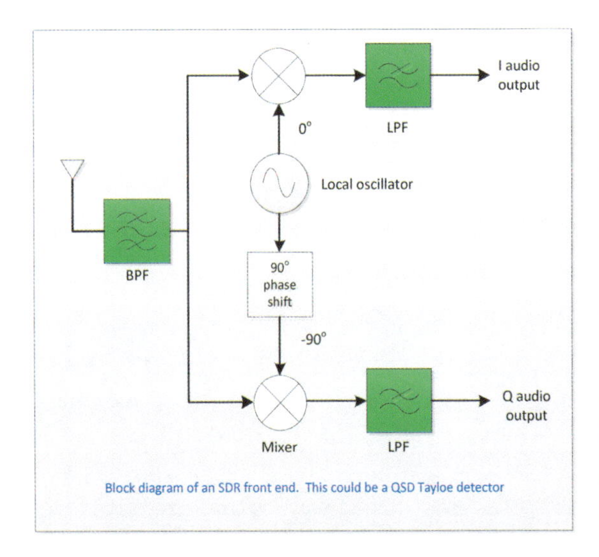

Block diagram of an SDR front end. This could be a QSD Tayloe detector

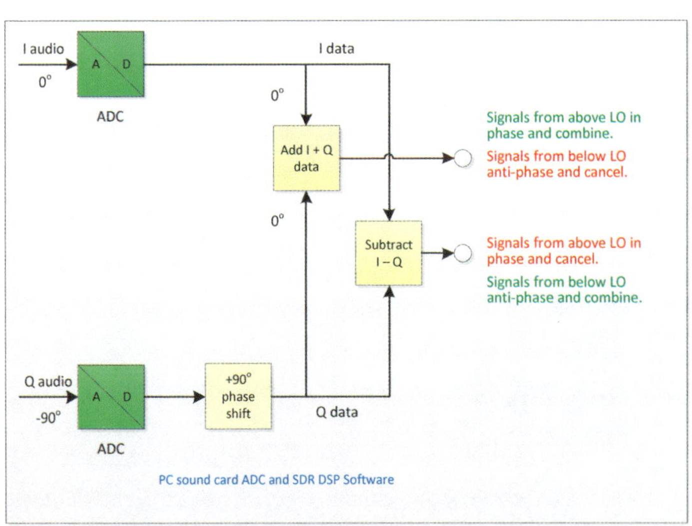

PC sound card ADC and SDR DSP Software

The diagram on the left shows the SDR front end, often a Tayloe detector. It is a direct conversion receiver like the direct conversion Superhet in the top drawing, with the addition of a second mixer and a 90 degree phase shift to generate the Q audio stream.

The diagram on the right shows the start of the DSP section. The two analog to digital converters might be the stereo PC sound card, or chips on the radio board. The yellow blocks are implemented in the PC software. They apply an additional 90 degree phase shift to the Q stream allowing recovery of the signals above the LO frequency with image frequencies cancelled out. Subtracting rather than adding the digital values allows recovery of the signals below the LO frequency with image frequencies cancelled out.

Vector diagrams of the image cancellation process

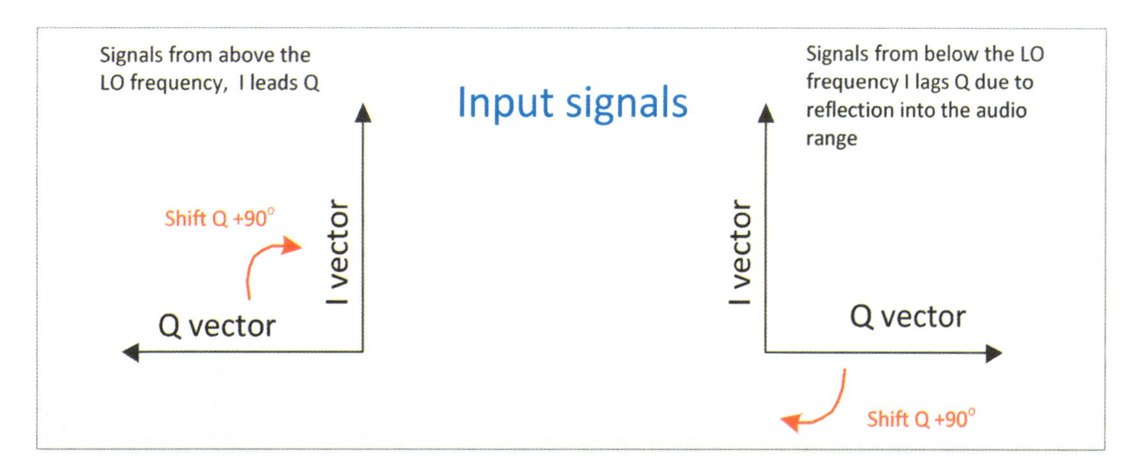

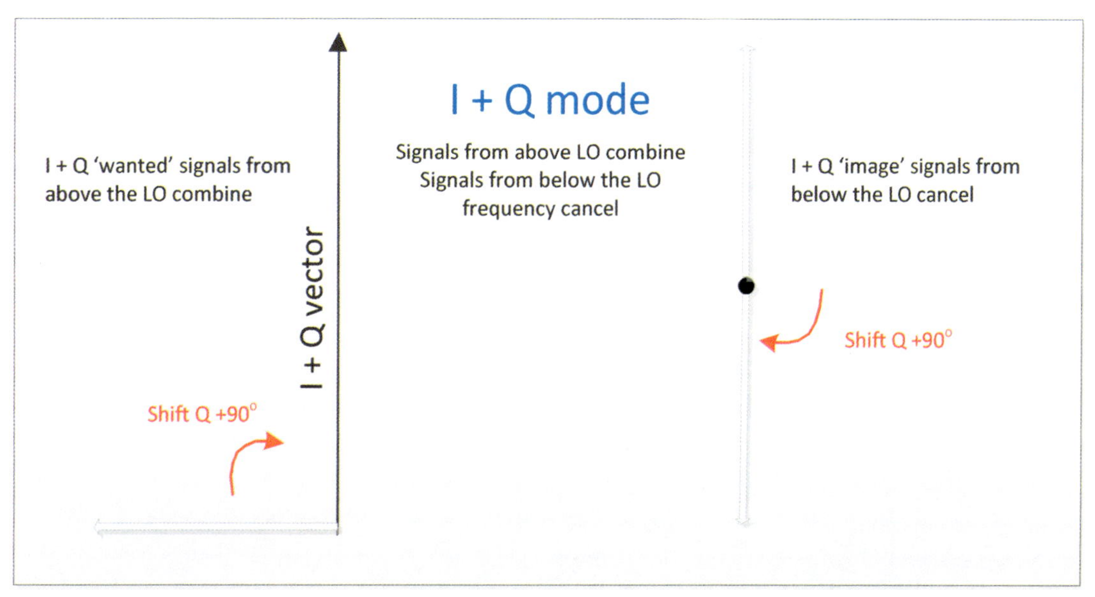

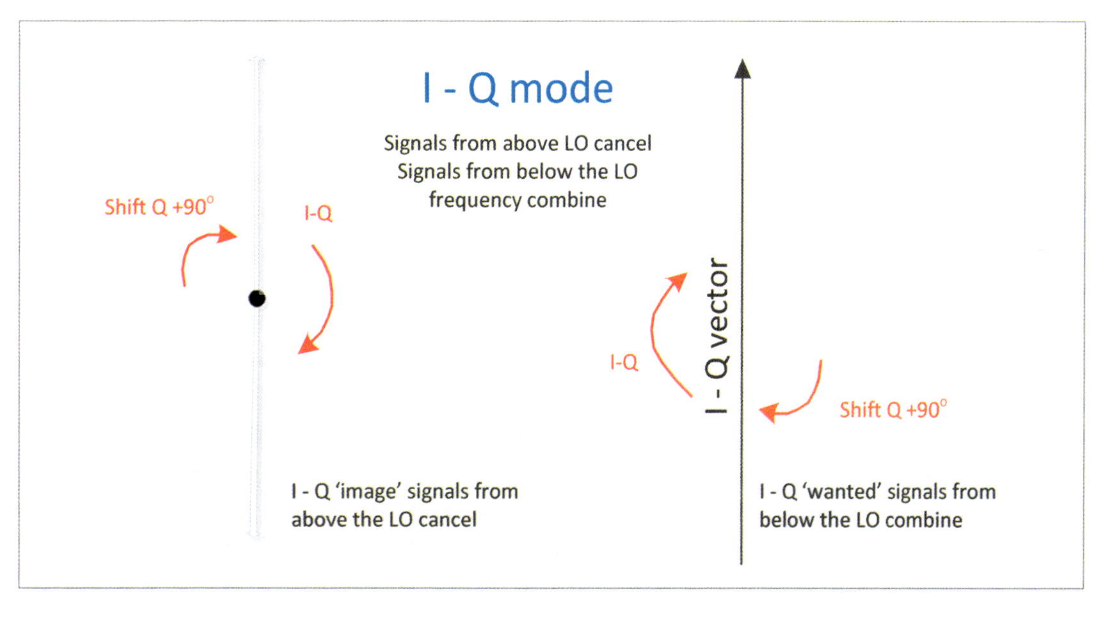

SDR Generations

The following table describes my interpretation of the definition of each SDR generation. Other interpretations exist, but this definition makes sense to me.

Generation	Definition
Generation 1 SDR	Software defined radio, usually employing a QSD (quadrature sampling detector) such as a Tayloe detector. It uses the sound card in a PC to perform the analog to digital conversion. Control, DSP, and demodulation are performed by PC software. If the device is capable of transmitting it will probably employ a QSE (quadrature sampling exciter) such as a Tayloe detector and will use the sound card in a PC to perform the digital to analog conversion. Control, DSP, and modulation are performed by PC software.
Generation 2 SDR	Software defined radio usually employing a QSD (quadrature sampling detector) such as a Tayloe detector. It uses an on board ADC (analog to digital converter) chip to perform the analog to digital conversion. Control, DSP, and demodulation are performed by PC software. If the device is capable of transmitting it will probably employ a QSE (quadrature sampling exciter) such as a Tayloe detector and use an on board DAC (digital to analog converter) to perform the digital to analog conversion. Control, DSP, and modulation are performed by PC software. VHF/UHF/SHF SDRs are often generation 2 SDRs.
Generation 3 SDR	Software defined radio, usually employing a QSD (quadrature sampling detector) such as a Tayloe detector. It uses an on board ADC (analog to digital converter) chip to perform the analog to digital conversion. DSP and demodulation is performed by chips in the radio. If the device is capable of transmitting it will employ a QSE (quadrature sampling exciter) such as a Tayloe detector or a conventional transmitter architecture. If it is a QSE it will use an on board DAC (digital to analog converter) to perform the digital to analog conversion. DSP and modulation is performed by chips in the radio. Many SDRs 'with knobs' which can work as stand-alone radios without connection to a PC are generation 3 SDRs.
Generation 4 SDR	Direct digital sampling SDR, using DDC (digital down conversion) in the receiver and DUC (digital up conversion) in the transmitter. The ADC and DAC are on board the radio and an FPGA is normally used to perform decimation and filtering which limits the data bandwidth requirement to the DSP stage which may be either on the radio circuit board or inside the PC software.

Chapter 3 - The DDC method

The last chapter dealt with the QSD / QSE quadrature sampling method which is used in most currently available SDR receivers and transceivers. Generation 1, 2 and 3 software defined radios all use the QSD method, normally employing a Tayloe detector. Generation 1 SDRs like the Softrock and many small SDR receivers use the PC sound card for analog to digital conversion and the PC for signal processing. Generation 2 SDRS such as the FLEX-1500, FLEX-3000 and FLEX-5000, use on board analog to digital converters and the PC for signal processing. Generation 3 SDRs use on board analog to digital conversion and have on board signal processing. The SDRs 'with knobs', which don't need a PC such as the Elecraft KX3, SDR Cube and ADAT ADT200A, are often generation 3 radios. This chapter is about 4th generation SDRs which use direct digital conversion.

I noted in the last chapter that a large part of the reason that SDR receivers perform so well is removal of the multiple mixers, local oscillators and IF amplifiers which make up a conventional superhetrodyne receiver. Eliminating these receiver components removes the noise and distortion they cause. In a QSD type SDR, the Tayloe switching detector is effectively a mixer so it can introduce some unwanted interference products and as the QSD is a direct conversion receiver it suffers from image signals which are not fully cancelled by the I/Q phasing method in the PC software. The bandwidth of 1st generation SDRs is limited by the sampling speed and performance of the PC sound card and all QSD receivers are limited by the bandwidth of the Tayloe detector. Most QSD receivers have pretty good performance and can display 48 kHz, 96 kHz or 192 kHz of spectrum. But now there is a new method which is even better!

The 4th generation of software defined radios use Direct Digital Sampling (DDS), specifically Direct Down Conversion (DDC) in the receivers and Direct Up Conversion (DUC) in the transmitters. DDC is the cutting edge of SDR design, offering some great new features and some significant improvements in performance. The claimed MDS (minimum discernible signal), intermodulation distortion, transmitter harmonic and sideband suppression figures for the new FlexRadio 6500 and 6700 transceivers are outstanding and will put them right at the top of the ARRL (QST) and Sherwood lists. I can't wait for those labs to publish their full test reports. Other DDS based HF transceivers include the Apache ANAN series, OpenHPSDR, hiQSDR, SunSDR2, and the ELAD FDM-DUO. HF DDC receivers include the Perseus, ELAD FDM-S1, QS1R, SDRIQ receivers and quite a few others.

In a DDC receiver the whole HF band is sampled more or less directly at the antenna and converted to a digital signal using a very fast ADC. Most models have front end filters before the ADC and some also have a variable attenuator and a low noise preamplifier for some or all bands.

We know that you need a sample rate of at least two times the highest frequency to accurately represent the RF spectrum. So for a receiver covering the HF and 6m bands, the ADC needs to sample at a minimum of 106 Msps, (106,000,000 samples per second). As an example, the new FLEX-6700 and FLEX-6500 cover from 30 kHz to 72 MHz using a sample rate of 245.76 Msps. The ANAN radios cover 10 kHz to 55 MHz with an ADC sample rate of 122.88 Msps. Once the whole HF spectrum has been sampled you can select the parts of it you want the software to display on the spectrum display.

Typically the software for these new high performance receivers can display between one and eight wideband receiver 'panadapters' at high resolution. Like other radios you can usually have two receivers on the same band allowing you to listen to both the DX station and the pile up when working split, but with an SDR you can also display and listen to several other bands at the same time. Some radios also allow more than two receivers on the same panadapter spectrum. You can place receiver panadapters in different windows, on multiple monitors and in some cases on multiple devices.

Sure it is neat, but initially I wondered about the usability of this amazing new capability. I can't think why I would want to listen to more than two signals at once, but viewing them is a different matter. Having seen receivers displaying up to eight bands simultaneously, I can see some benefits. Firstly you get an appreciation for which bands are open if you can see the activity on all of them at once. Other possibilities are monitoring a net or sked frequency on one band while looking around or working stations on other bands. Have you ever lost a QSO when attempting to move to another band? Maybe the frequency you chose is already in use or you can't hear the other station when you QSY. With multiple receiver panadapters you can move to the new frequency and still monitor the old one. It is going to be a real advantage for contesters, they will be able to see band conditions improving on one band and decreasing on others. No more wondering if 10m is open but being unwilling to move off your spot on 20m. Some of the new radios can assign different antenna inputs to different bands or use two antennas simultaneously for diversity reception. You could use this to combine the signal from a receive only antenna like a Beverage with the signal received on your transmit aerial. Or perhaps you could combine signals from two antennas, one facing Europe and the other facing Asia or Oceania.

On the transmit side, the analog signal from your microphone or digital mode software is immediately converted to a digital signal, modulated and processed in the software. The RF signal is generated directly at the transmit frequency using a DAC. It is not mixed up to the RF frequency from a lower IF frequency, like QSE SDRs and conventional transceivers. This is known as DUC (digital up conversion). Like all digital technology the advantage of digital up conversion is that digital signals can be translated in frequency and filtered without the introduction of noise. In some SDR transceivers the CW signals are generated directly at the transmit frequency rather than as an audio tone at the beginning of the transmit chain. In other words there is no modulator for CW, the RF signal is simply turned on and off or more accurately the RF signal is either generated or not generated, creating a very clean CW signal with fast QSK switching. Another new technique under development is to use pre-distortion to dynamically correct for non linearity in the RF power amplifier. This can result in an even better transmit signal with lower harmonic and intermodulation products.

One of the most exciting developments in DDC technology is the recent release of three new 100W transceivers and a receiver from FlexRadio Systems and the release of four new transceiver models from Apache Labs. The FlexRadio 6700 transceiver and FlexRadio 6700R receiver each allow the creation of up to 8 independent Slice Receivers on up to 8 panadapters, providing reception from 30 kHz to 72 MHz and 135 MHz to 165 MHz. The FlexRadio 6500 allows creation of up to 4 Slice Receivers on up to 4 panadapters with tuning from 30 kHz to 72 MHz. Each panadapter can be up to 14 MHz wide. The FlexRadio 6300 covers 30 kHz to 55 MHz and can support two 7 MHz wide panadapters and 2 receivers.

Quoted performance figures for the FLEX-6500 and FLEX-6700 include; receiver spurious and image rejection better than 100 dB, and local oscillator phase noise performance better than -147 dBc/Hz at 10 kHz offset and -152 dBc/Hz at 100 kHz offset. Transmit carrier and unwanted sideband suppression better than 80 dBc and harmonic suppression better than 60 dBc. [Source FLEX-6000 Family Datasheet.pdf © FlexRadio Systems].

Apache Labs has released the ANAN-10 and ANAN-100 (10W and 100W) transceivers which are based on the Hermes SDR transceiver board produced by volunteer members of the 'OpenHPSDR' group. *"The HPSDR is an open source (GNU type) hardware and software project intended as a "next generation" Software Defined Radio (SDR) for use by Radio Amateurs ("hams") and Short Wave Listeners (SWLs). It is being designed and developed by a group of SDR enthusiasts with representation from interested experimenters worldwide"*, see (http://openhpsdr.org/). Apache also offers the ANAN-100D based on the Angelia board which is an updated Hermes board with a much faster FPGA and twin analog to digital converters, and the ANAN-200D based on the even more powerful Orion board.

The four Apache Labs transceivers cover 10 kHz to 55 MHz and can display up to seven receiver panadapters 384 kHz wide. The PowerSDR mRX SDR software can combine several panadapters on the same band to increase the displayed bandwidth on the primary panadapter to around 1.1 MHz.

Apache quotes an MDS of -138 dBm, receiver image and spurious response rejection better than 100 dB. A blocking dynamic range of 125 dBm and local oscillator phase noise performance better than -149 dBc/Hz at a 10 kHz offset. The transmit carrier and unwanted sideband suppression is stated to be better than 90 dBc and harmonic suppression better than 50 dB. [Source 'ANAN RADIOS SPECS & INFO' © Apache Labs].

The FlexRadio 6700 and 6500 have a 100W full duty cycle power amplifier and an internal antenna tuner. The ANAN-100 radios have a 100W PEP power amplifier and do not include an internal antenna tuner. The FLEX-6700 can use its two ADCs, FlexRadio calls them 'Spectral Capture Units', as a full diversity receiver. The ANAN-100D and ANAN-200D can use their two ADCs in the same way. The ANAN-200D has an option for a 3rd ADC.

These new transceivers use 1 Gbps Ethernet ports to connect to the PC. But there is a difference in both the technical design philosophy and the overall sales approach taken by these two leading companies. First the technical approach; the Flex radios follow the trend for moving as much processing power as possible back into the radios. Their on-board Vitex-6 FPGA and DaVinci DSP chips do all the signal processing including modulation, demodulation and filtering. The PC software is only used for display of the receiver panadapters and control of the radio. Because the demand for data over the connection between the radio and the computer is much reduced, this type of connection is called a 'thin client' application. The Apache radios follow the traditional SDR method where much of the signal processing including modulation, demodulation and filtering is done by software in the PC, in addition to the display and control functions. As there is a need for a continuous high speed data connection between the radio and the computer, this is known as a 'thick client' application.

There are pros and cons in both approaches. The thin client approach reduces the load on the PC and allows for the use of Netbook, and Tablet type computers connected to the network via WiFi. It will eventually make remote operation over the Internet easier, but this functionality is not available for the FlexRadio yet. The thick client approach needs a more powerful PC but is cheaper to produce making the radios significantly cheaper. It is easier and much more efficient in terms of code lines to write code for the PC application than Verilog code for the FPGA, so thick client radios are easier and faster to develop. This makes the thick client model more suitable for open source projects and several software packages may be available for the same radio. On the other hand, once a thin client radio has been developed, writing PC software to control it is easier since there is no requirement for complex FFT math and DSP functions like dynamic filtering, modulation and demodulation. Whether thick client or thin client is better is yet to be determined.

The two companies also have different approaches to their product development. The Apache radios are built in collaboration with the OpenHPSDR group which is a group of volunteer developers dedicated to the open source development of SDR software and hardware. When you buy an Apache radio you are only buying the transceiver hardware. The required software can be downloaded for free, but it is not supported by Apache Labs. Software for the ANAN radios is written by various software developers usually under the 'open source rules'. Currently there are several compatible open source free applications; PowerSDR mRX, Kiss Konsole, ghpsdr3, Hetrodyne (MAC) and cuSDR (receive only). The ANAN series can also be used on receive only with Studio 1 which is commercial software written by SDR Applications and distributed by WoodBox Radio.

This marriage between a commercial hardware product and open source developed software used to be followed by FlexRadio as well. Initially their PowerSDR software was open source and available to any software developers to modify for their own use on the basis that it remained a free, non-profit product. When the ingenious multiple tracking notch filters function was added, FlexRadio retained propriety rights to that code although they continue to allow free unrestricted download of their PowerSDR software.

With the introduction of the new Signature 6000 series, FlexRadio Systems has decided to move away from the old approach and their new radios will only work with the FlexRadio SmartSDR software shipped with the radios. There is a charge for updates (but not bug fix upgrades) after the first year of ownership. Given that the software is such a major component of the SDR radio and considering the high cost of the software development, charging for upgrades is understandable. Some other SDR software, both open source and propriety, is optimised for particular SDR hardware. It can be very confusing finding out what software will work with which hardware.

The Apache radios and many SDR receivers use open source software, where the software is developed by amateur developers for free and there is no guarantee of any future support or bug fixes. Surprisingly, given the challenges for developers who have only their leisure time for programming, the open source software such as PowerSDR mRX, cuSDR, SDR#, HDSDR and many, many, others is extremely well written and performs very well indeed. In most cases the developers seem to be very approachable, interested in extending the software and responsive to constructive criticism and requests.

Chapter 4 – SDR performance measurement

When I started this book, one of the questions I asked, was; "Do SDRs perform better than conventional Superhet architecture radios?" There is no clear answer to my question because there is a wide variance in the performance of both conventional and SDR radios. Some of it is related to the price you pay, the internal architecture of the radio, quality of components, and how recent it is. Some of the recent offerings from the 'big three' equipment suppliers have much better performance statistics than the radios they released ten years ago. SDR performance is affected by the PC and the SDR software as well as the radio itself.

Based on the numbers published in QST lab tests, conventional superhet receivers do sometimes outperform SDR receivers. But if you compare radios at a similar price the SDR radio is likely to beat the conventional radio in some of the key tests. In the real world with many signals on the band, the SDR will usually have superior IMD (intermodulation distortion) performance because, within limitations, the IMD performance of an SDR improves when there are multiple signals.

My own list, ranking 30 receivers, is based on an average of the receiver tests in 27 QST reviews and online data for three other radios. SDR receivers hold 1st, 2nd, 6th and 9th place and there are seven SDR receivers in the top 20. There are very few 100W ham radio SDR transceivers on the market. My ranking of transmitter performance, including QRP transmitters, places three SDRs in the top 10 and five in the top 20. Currently the data available for the FlexRadio 6700 is rather limited. The few measurements available from FlexRadio place it 2nd on the receiver table, behind the ANAN-100D, and 1st on the transmitter table, but the rankings might change when QST, Radcom or other lab tests are published.

A big problem is that several of the tests traditionally used to compare radios on league tables like the Sherwood Engineering list and in Radcom and QST reports are not very relevant to SDRs because of the fundamental differences in technology. For example the Sherwood Engineering list is sorted on the results of the, 'narrow spaced two tone IMD' performance test. This is probably because that test indicates how well a receiver would perform when the band is busy such as during a contest. The problem in using this approach when comparing SDR receivers with traditional Superhetrodyne receivers is that the causes and effects of intermodulation distortion is completely different in SDR receivers. SDR receivers often look very good on the two tone IMD test so they often rank quite high on the Sherwood list and in reviews, but the traditional test method can unfairly penalise SDR receivers.

When we are making a decision about which transceiver to buy, or add to our 'Lottery list', we want to make an informed decision so we check out the online reviews, advertisements and technical reviews. A big part of most reviews deal with the functional aspects such as; what features are included, ergonomics, and how well they work on the air. These are very important issues and in many cases they should be the major factors in deciding whether the rig is right for you. After all you want a radio which suits the way you operate. I prefer to use SDR radios because I really like the spectrum display and 'click on the signal' operation. Another very important thing for many of us, is the balance of what you get versus the cost for the unit, i.e. "How much bang for your buck."

The reviews published in QST magazine and Radcom include test results from their lab testing. Other great resources are the test results published on the Sherwood Engineering web site and the Adam Farson VA7OJ / AB4OJ web site.

I see many comments online declaring that the test results mean nothing and the only way to evaluate a radio is by using it on air. I disagree. The tests are designed to simulate conditions found in real world situations, like a contest weekend, rag chewing on 80m, or another ham in the same street operating a little further up the band. The advantage with lab test results is they are repeatable and they offer an unbiased comparison between radios. You just need to know how to read the numbers and work out which tests are the most important considering the way you want to operate. For example, contesters might want different performance than those who want to work weak signals on 160m, or EME on 6m. You also need to understand that SDRs are fundamentally different and different tests need to be applied to them.

Until recently I was completely unaware that some of the traditional tests of receiver performance are not valid for Software Defined Radios. Adam Farson VA7OJ/AB4OJ has done a lot of work on testing conventional and SDR radios and has explained the issues. He has also introduced a new Noise Power Ratio test which I believe is a really good performance indicator for radios with wide band front ends like SDR receivers.

All of the common transmitter tests such as RF power, carrier and unwanted sideband suppression, two tone IMD, transmitter harmonics, and composite noise, are just as important in SDR transmitters as they are in conventional ones.

Receiver **minimum discernible signal** (MDS) is a measurement of the weakest signal which can be heard in the receiver, i.e. 1 dB above the noise floor. In most cases when you connect your antenna to the radio the noise level rises, indicating that the background noise being received at your QTH is higher than the noise generated inside the receiver. So having very good sensitivity may not be as important to you as the other tests. But having a good MDS figure would be important for receiving weak signals when the band is quiet, or on 160m when using a low gain antenna like a Beverage.

The **3rd order IMD dynamic range** (DR3) is completely different in an SDR and the practice of quoting a 2 kHz offset DR3, measured when the intermodulation products become equal to the MDS level is not correct for SDR receivers. In a superhetrodyne receiver the non-linear process of mixing causes intermodulation distortion (IMD) products. Large signals near the receive frequency can cause interference on your receive frequency. This is a big problem on contest weekends when you want to hear weak signals operating on frequencies close to stronger stations. Direct sampling SDRs do not have mixers (in hardware) so they do not suffer from the problem of mixers causing IMD, however the ADC does cause some intermodulation distortion so the test is still useful if it is measured correctly, preferably with the results presented on a graph. The traditional test method measures the input level of the two tone test signal when intermodulation products become noticeable on the receive frequency, i.e. when the IMD level is equal to the MDS level. This is a 'best case' scenario because as the two tone level is increased further, the IMD dynamic range degrades in a linear fashion.

In an SDR the IMD dynamic range does not degrade in a linear fashion, it stays the about same when the two tone test level is increased and in most cases actually improves until the input level gets fairly near the clipping level of the receiver. When the test signal level reaches the receiver clipping level the IMD performance crashes. In the real world this means a conventional receiver will suffer worse IMD distortion in the presence of several medium size signals like a contest going on, than an SDR will. The IMD performance of an SDR actually gets better when the band is busy! The 3rd order IMD dynamic range of an SDR should not be measured at the MDS level, but at the level where there is the greatest difference between the two tone input level and the IMD product level. This has been known for a long time, but most labs still measure at the MDS point, which may not give a fair result on an SDR. To quote Leif Åsbrink, SM5BSZ in QEX Nov/Dec 2006, *"Rather than measuring what level is required for getting IM3 equal to the noise floor, one should measure the largest difference (in dB) between the test tones and the intermodulation product. It will be close to saturation of the A/D converter on the SDR-14, while it will be at the noise floor for an analog receiver."*

The often quoted **3rd order intercept point** which is never measured directly anyway can't be measured for an SDR receiver because the relationship between the IMD products and the input tone levels is not linear. The two lines diverge rather than crossing, so there is no IP3 point. This does not stop many SDR manufactures quoting an IP3 using a calculation based on the 3rd order dynamic range.

The **Reciprocal Mixing Dynamic Range** (RMDR) test measures how well a receiver can cope with a single large signal just outside the pass band you are listening to. I call it a, "Ham next door test," because it demonstrates how well your receiver works when the ham down the street transmits near your operating frequency. In a conventional receiver, reciprocal mixing noise is caused when noise from the LO (local oscillator) mixes with strong adjacent signals, generating noise at the output of the mixer. The generated noise can degrade a receiver's sensitivity and it is most noticeable when the offending signal is close to your receiver frequency, which is why RMDR is usually reported at a close offset of 2 kHz and at a wider offset of 20 kHz.

In a direct sampling SDR receiver reciprocal mixing noise is caused when phase noise from the ADC (analog to digital converter) mixes with the offending signal, so RMDR is an indicator of the spectral purity of the ADC clock. In an SDR the RMDR is usually completely independent of the offset from the receiver frequency and is normally the same at 2 kHz and at 20 kHz.

Noise Power Ratio can be measured in either type of receiver and is a very good test of SDRs because they are wideband receivers. The idea of the NPR test is to load up the receiver with lots of signals. This is simulated by connecting a 2, 4 or 8 MHz wide 'white' noise source to the receiver input. You increase the noise level until the signal is slightly under the clipping level. Then you cut a slot in the noise exactly 3 kHz wide using a very good band stop filter and measure the amount of noise caused by intermodulation products generated in the receiver which fall into the narrow quiet spot. Adam Farson measures the level in a 2.4 kHz wide bandwidth in the bottom of the slot since this is a typical bandwidth for SSB and it fits neatly inside the 3 kHz slot. The NPR is the ratio of the noise power inside the slot to the level of noise power in a channel the same bandwidth outside the slot. Since the SDR receiver has a spectrum display the result can be read straight off the screen.

A 4 MHz wide noise signal at just under clipping level is equivalent to more than 1200 SSB signals at S9 +30 dB. This is the ultimate test of how well the SDR will perform in a busy band.

Receiver dynamic range is the receiver's ability to handle a range from weak signals near the MDS level to very strong signals like S9 +60 dB. In SDR receivers the number of bits the ADC uses and the decimation ratio, affects the dynamic range. But compared to a conventional receiver it is still quite large. Conventional superhetrodyne receivers use AGC to extend their dynamic range, but this is not necessary on SDRs which typically have around 120 dB of dynamic range in a 2.4 kHz bandwidth. There is a great deal of misinformation on the Internet about this, but that discussion will have to wait for another chapter. One downside of SDR receivers, is that the dynamic range is affected by the sum power of all signals within the 0 – 60 MHz bandwidth of the receiver. If there are many large signals, the dynamic range will shrink. This can be demonstrated using the Noise Power test set which simulates a high loading of signals. In a test performed by Adam Farson a single tone injected into a FLEX-6700 demonstrated a clipping level in excess of +13 dBm, but a worst case white noise signal 8 MHz wide from the NPR test set reduced the clipping level to -2 dBm. This represents a 15 dB decrease in dynamic range, so it is lucky the receiver has a very large dynamic range to begin with. Of course this level of reduction could never happen in the real world because the loading could never approach the white noise level. Eight bit SDRs which have a limited dynamic range and don't have filters to reject the AM broadcast band will suffer degraded dynamic range due to this effect.

In the **Blocking Dynamic Range** test a wanted signal on the receiver frequency and an unwanted signal at either 2 kHz or 20 kHz offset is input to the receiver. The unwanted signal is increased until the wanted signal measured at the receiver output decreases in level by 1 dB. SDRs don't normally suffer from blocking de-sensitivity. The test 'unwanted' signal normally reaches clipping level with no effect on the wanted signal. So BDR can be measured for an SDR but it may not mean much since you just record the ADC clipping level.

BDR is sometimes stated as the input level when the noise floor in a 2.4 kHz bandwidth at the wanted frequency rises by 1 dB. Instead of measuring a decrease in the wanted signal level, an increase in noise within the 2.4 kHz bandwidth is measured. This method assumes that a 1 dB increase in the noise level indicates that the signal to noise level of the wanted signal, if it was present, would have been reduced by 1 dB. A 1 dB decrease in the signal to noise ratio is considered to be equivalent to a 1 dB reduction of the wanted signal. BDR tested this way can be measured for an SDR due to the dynamic range compression noted above and the method is easier to perform since you only need one test tone.

Image rejection is not very relevant to DDC SDRs since they don't have hardware mixers. But you can test image rejection in QSD type SDRs which due to being direct conversion receivers are prone to image signals. A better test for direct down conversion SDRs would be to check the filtering used to prevent 'aliased' signals appearing in the receiver pass band. ADCs usually have a much wider frequency response than the Nyquist bandwidth related to the sampling frequency. For example if the sampling rate is 122.88 MHz, as used in the ANAN / Hermes receiver, the Nyquist bandwidth extends up to 61.44 MHz. The 2nd Nyquist zone is from 61.44 - 122.88 MHz and the 3rd Nyquist zone is from 122.88 to 184.32 MHz.

The LTC2208 ADC used in the ANAN / Hermes receiver has a frequency response up to 700 MHz, so without filters ahead of the ADC, large signals in the 2nd to 11th Nyquist zones could be sampled by the ADC. Once sampled, they would end up being heard in the receiver and shown on the spectrum display. You will note that the 2nd Nyquist zone includes the FM broadcast band. The ADC is designed with a wide frequency response so that developers can deliberately use the alias effect to receive frequencies well above the sampling frequency. This is called under sampling. In the Hermes design, a 55 MHz low pass filter is incorporated to attenuate alias signals and ensure they do not get sampled by the ADC.

As the signal is decimated inside the FPGA from 61.44 MHz down to the 48 kHz - 384 kHz panadapters, additional alias zones are created. The FPGA software includes CIC and CFIR filters to ensure any signals from these new alias zones are attenuated by at least 100 dB and will end up below the receiver's noise floor.

Chapter 5 – The SDR transceiver, what's in the box?

This chapter is about the building blocks making up a DDC SDR transceiver. What are the components, why are they used and how do they work? Much of the work is done inside the FPGA so I will discuss both the hardware design of the radio and the blocks of software coding inside the FPGA. I will use the Apache Labs ANAN-100 as an example because it is based on an open source design and there is a lot of information available about how it works. I am definitely not an expert in SDR design. I have very little programming experience and have not been involved in any SDR development. Like you I am interested in learning about SDRs, so I am using this exercise to research and understand how the radio works.

The ANAN-100 transceiver is based on the very successful Hermes and Alex (Alexiares) boards developed by the OpenHPSDR group (http://openhpsdr.org/), plus a 100W power amplifier developed by Apache Labs in association with the OpenHPSDR group. In turn, the Hermes board is based on three earlier Open HPSDR projects; the Pennylane transmitter, the Mercury receiver and the Metis Ethernet interface. The ANAN-100D uses an Angelia board which is a completely new board based on the Hermes design, with two ADC chips and a much larger and faster FPGA. The ANAN-200D is based on the even more advanced Orion board. Most of the following description of the ANAN-100 is also valid for the ANAN-100D, the ANAN-200D and in principle other DDC SDR receivers and transceivers.

The block diagram for the ANAN PA/Filter board is included in the ANAN-100 user guide which can be downloaded from the Apache Labs web site. Or download 'ANAN RADIOS SPECS & INFO' from the Apache Labs instant download area. A block diagram of the Hermes transceiver is shown below.

In 2008 the annual ARRL and TAPR Digital Communications Conference included a seminar by Phil Harman VK6APH who was the project leader for the development of the Mercury receiver. The conference was recorded and the DVDs are available from www.arvideonews.com/dcc2008/. The Phil Harman talk is also viewable on YouTube as Ham Radio Now - Episode 62 parts 1, 2 and 3. Each part is an hour long but they are well-paced, entertaining, not too technical and very highly recommended. The three videos are the best description of the current DDC SDR technology I have found. This chapter is based on his talk, as well as Apache and OpenHPSDR documents downloaded from the Internet.

The Power Amplifier and Filter board

The normal receive path is via one of the three main antenna connectors. There are also two antenna ports for receive antennas, a transverter port, and a bypass output port designed so you can insert an external filter or a preamplifier. The normal signal path is into the low pass filter section which is used on both transmit and receive. There is a choice of seven software selectable low pass filters for the ham bands. The LPF filter section is not in circuit if the EXT1, EXT2 or RX2 antenna inputs are used. Neither the LPF nor the Alex HPF is in circuit when you use the RX2 input on an ANAN-100D or 200D which have a 2nd ADC.

On receive; the output of the low pass filters is routed via the transmit / receive (TR) switch to the high pass filter section, which contains five software selectable high pass filters, that can be bypassed if required. The five high pass filters are not ham band specific, they break the HF spectrum into zones so the receiver retains general coverage receive capability. The sixth filter selection is a band pass filter for the 6m band and includes a low noise amplifier. The board also includes a software selectable 10 / 20 / 30 dB attenuator before the high pass filters.

Why do we need all these filters?

There seems to be a lot of filters in the receive path. Even if you bypass all of the filters on the PA/Filter board, there is still a Mini-circuits 55 MHz low pass filter at the input of the Hermes transceiver board. Why does the radio need all of these filters? There are two answers and one is much more important than the other. The less important reason is the dynamic range of the ADC is affected by the total power of all the signals presented to it. This includes signals within the Nyquist bandwidth such as AM broadcast stations and also signals within multiples of the Nyquist bandwidth, right up to 700 MHz in the case of the LTC2208 ADC in an ANAN receiver. Receiving a lot of strong signals reduces the dynamic range of the receiver and the displayed noise floor rises.

But the more important reason for employing filters is to ensure that the ADC does not sample frequencies from above the HF band. The same 'out of band' signals that can cause dynamic range compression could be sampled, causing interference with the wanted 'in band' signals. The low pass filters block 'out of band' signals preventing breakthrough of signals from above 55 MHz including FM broadcast and VHF TV stations. The high pass filters block unwanted signals below your wanted range such as AM broadcast stations, ensuring there is minimal reduction of the dynamic range and signal to noise ratio. A medium level of signals within the Nyquist bandwidth, between 0 and 61.44 MHz, will not affect the noise floor much and will act to reduce the effects of intermodulation distortion, so the 'in band' unwanted signals should be reduced, but they do not have to be entirely eliminated.

The ability to sample frequencies above the Nyquist bandwidth is known as aliasing and is used in commercial SDR receivers for receiving signals above the sample rate. In our amateur radio transceiver, we don't want interfering signals from above the Nyquist bandwidth because once they are sampled they can't be removed. The 55 MHz low pass filters ensure any aliased signals are attenuated by at least 100 dB and end up below the noise floor. Phil Harman said that he can't see any FM broadcast signals on his Mercury receiver even without the Alex filters, so I believe this issue has been fully resolved.

The FM broadcast band extends from about 88 MHz to 108 MHZ, which due to aliasing would appear as signals between 26.56 and 46.56 MHz in the receiver. At my QTH I cannot see any sign of the local FM stations even when the filters on the PA/Filter board are set to bypass.

Using the front end filters messes up being to take advantage of the radio's ability to display multiple receiver panadapters on different bands and the wideband 0-55 MHz spectrum display. Depending on the band choice, you may have to bypass the filters if you want to view panadapters on multiple bands. In most cases, unless you live in Europe where you may live close to FM and TV stations, bypassing the filters will have no effect on the receiver performance. At my QTH, I see no increase in the noise floor and no unwanted new signals when the filters are switched to bypass mode. It is difficult to persuade PowerSDR to bypass the low pass filters because they are automatically selected for the transmit frequency in use. You can adjust the frequencies at which it changes filters, so the LPF switches to bypass when the radio is tuned to a frequency outside of the ham bands, but the default condition is for a low pass filter to always be in circuit. If you want full unfiltered receive capability it is probably easier to use the EXT1 or EXT2 antenna connectors. With the cuSDR SDR software, the LPF engages for each ham band and automatically drops to bypass when the radio is tuned to frequencies outside of the ham bands.

The Hermes board

The input to a Hermes SDR board can be connected directly to an antenna, but in the ANAN-100 it is connected via the PA/Filter board. The Hermes receive antenna connects to a Mini-Circuits software selectable 0 – 31 dB attenuator and then a 55 MHz low pass filter, followed by a 20 dB low noise preamplifier, which is always on. By setting the normal condition of the adjustable 20, 10, or 0 dB, attenuator on the PA/Filter board to 20 dB of attenuation, the combination of the attenuator and the amplifier works like a preamp stage with 0, 10 or 20 dB of gain.

The ADC overload point of an ANAN-100 is about +8 dBm with the preamp "off" (attenuator at 20 dB) and -12 dBm with it "on", (attenuator at 0 dB). The MDS in a 500 kHz bandwidth is about -118 dBm with the preamp "off" and -138 dBm with it "on." It is very important that the low noise amplifier should not degrade the noise figure or intermodulation performance of the radio. The preamp used in the ANAN / Hermes/ Angelia / Mercury receiver is a Linear Technology LTC6400 amplifier specifically designed for use with the matching ADC.

The next stage of the Hermes receiver is the ADC. The Linear Technology LTC2208 is about the best choice that the designers had available at the time at a reasonable cost. It is a 16 bit ADC with a good SINAD (signal to noise plus distortion) ratio, and optional internal dither and random functions. The PC software has the option of turning dither and random on or off.

As a comparison, the FLEX-6000 series of radios use the Analog Devices AD9467 16 bit ADC and clock it at a higher rate so the radio can cover up to 72 MHz (and VHF on the 6700). It has a very slightly (about 1 dB) worse signal to noise rating, resulting in an ENOB of 12.1 bits vs 12.6 calculated for the LTC2208. It also does not have internal dither and random functions, which I personally feel is no big deal since I don't use those features anyway. The wider frequency coverage and higher sample rate make up for the very minimal difference in noise performance. Both ADCs have the same 100 dB SFDR dynamic range.

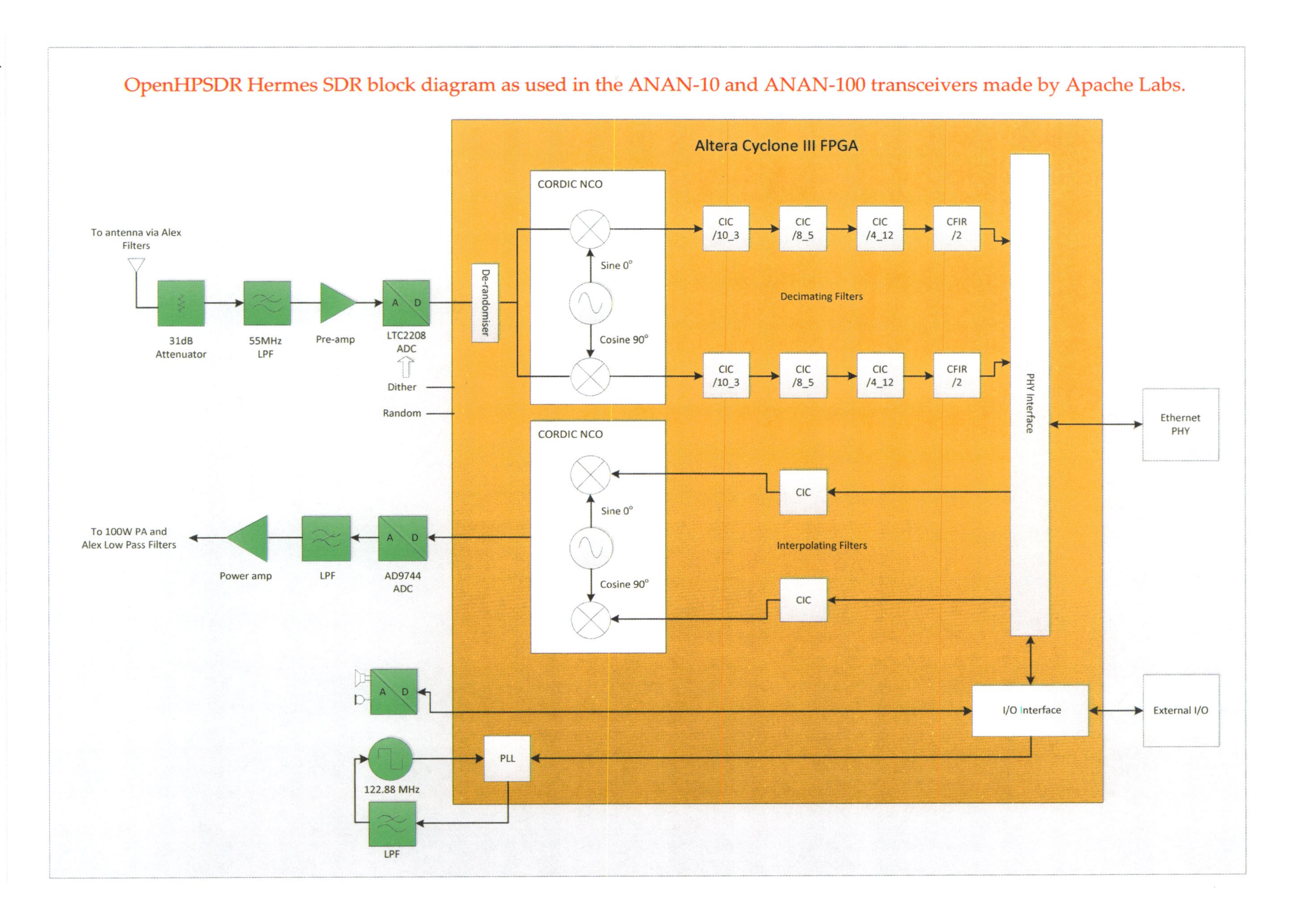

OpenHIPSDR Hermes SDR block diagram as used in the ANAN-10 and ANAN-100 transceivers made by Apache Labs.

Do I need to dither?

Dither improves the intermodulation performance of the ADC by ensuring the input signals are well spread across the analog to digital transfer characteristic of the ADC. Noise is added to the analog signal and then removed after the analog to digital conversion. In the LTC2208 ADC the noise is generated as a pseudorandom digital code inside the ADC. The code is converted into an analog signal and mixed with the incoming signal from the antenna. After the combined signal has been sampled and converted to digital data the original digital code is mathematically subtracted back out again. The result is better IMD performance with only a small degradation in noise performance.

The signal the ADC samples is the vector sum of everything at its input, so unless you have a particularly poor antenna there will usually be enough medium size signals to ensure that the ADC input signal will transverse most of the ADCs range. Dither could be helpful if you are using external narrow band filters for contest operation and it does improve the IMD performance when a two tone test signal is used as part of lab testing. In the real world with the antenna connected there are many signals being processed by the ADC so using the dither function is probably not necessary.

What does the random function do?

"The ADC is a small device with a big problem." On the antenna side there are very small sub microvolt signals which we want to hear in our receiver and on the other side there are very high speed 3.3 V square waveform data and clock streams. Inside and around the chip both signals are in very close proximity. The ADC samples at 122.88 Msps and the same clock drives the FPGA. At 16 bits per sample the output stream is a square wave at around 1.97 GHz. Capacitive and radiated RF coupling of the large 3.3 V signal into the sensitive analog input stage is a problem even with very good circuit board design. The effect, particularly when there are a limited number of input signals such as during testing, is that the coupling causes fixed signal spikes called spurs which are visible on the spectrum display and can be demodulated. The generation of spurs is a particular problem when the radio is being used in a situation where the receiver signal(s) are at fixed levels or frequencies such as a fixed radio link or cellular base station. The random nature of the signals on the ham bands means the input signal is fairly random and so the output signal is quite variable. In most cases this means the internal random function does not need to be used while listening on the ham bands.

The problem with fixed levels and signals on the ADC input is the repetitive nature of the data coming out of the ADC. This causes harmonics and related intermodulation products which could be coupled into the ADC input. Signals up to 700 MHz can be aliased back into the wanted bandwidth and they are being generated after our front end filters. The answer is to randomise the data coming out of the ADC so any coupling becomes noise rather than visible and audible spurs. The LTC2208 ADC does this by applying an Exclusive OR function to bits 1 to 15 of the output signal based on the value of bit 0. This produces a randomised output bit stream resulting in a clean spectrum display with no spurs. As soon as the scrambled signal gets to the FPGA it can be converted back into usable data by simply applying another Exclusive OR function to bits 1 to 15 based on the value of bit 0.

The output of bit 0 is random due to the noise performance of the ADC so the scrambling function really does produce a randomised output. ADCs used in other SDR receivers may not include the internal application of dither and random. The functions may be added with other hardware or not used at all.

On my system turning dither on degrades the noise floor, in a 2.4 kHz or 500 Hz bandwidth, as indicated by the S meter by between 2.5 and 3.5 dB. If dither is turned on, turning random on or off has no noticeable additional effect on the noise level. If dither is off, turning random off as well improves the noise level by around 0.8 dB. Since the higher noise level with dither on is still much lower than the noise floor with the antenna connected, this minor increase in the noise floor is fine. It does not affect the signal to noise ratio of received signals. With the antenna connected I can still sometimes see and hear a very slight increase in noise when dither is turned on. I can't see any change when random is turned on. These two settings are a personal preference, but since I don't see any distortion or spurs when I turn random and dither off, I usually leave both disabled.

The FPGA

Now the signals have been converted into a digital data stream they can be passed on to the DSP part of the radio inside the FPGA. The functions of the FPGA can be described in terms of analog signals and depicted as hardware gates and devices but, like a computer CPU, all of the signal processing is achieved by the manipulation of digital bits. The internal gates, registers and memory are built out of simple logic elements and programmed connections rather than existing as physical devices.

The Altera Cyclone III or Cyclone IV FPGA in an ANAN radio performs the tasks of down conversion and filtering. It does this job very well and very fast. There is plenty of spare capacity in the FPGA which could be used for additional functionality in the future. For example, the Cyclone IV FPGA chips in the ANAN-100D and 200D are currently only around 30% utilised. Some FPGAs can be configured to act like a 'software' microprocessor CPU, allowing them to run programs. This could be used to control functions in the radio. The FPGA chips in ANAN SDRs have between 39,600 and 150,000 logic elements which can be configured to do many tasks simultaneously. An FPGA may be a little slower than a microprocessor but it has a much higher capacity for multi-processing and it can be reconfigured internally at any time.

A field-programmable gate array (FPGA) is a semiconductor device that can be programmed after manufacturing. Instead of being restricted to any predetermined hardware function, an FPGA allows you to program product features and functions, adapt to new standards, and reconfigure hardware for specific applications even after the product has been installed in the radio. An FPGA contains programmable logic components called logic elements and a hierarchy of reconfigurable interconnects that allow the logic elements to be physically interconnected. You can configure logic elements to perform complex combinational functions, or merely simple logic gates like AND and XOR. In most FPGAs, the logic blocks also include memory elements, which may be simple flip-flops or more complete blocks of memory.

Like a microprocessor, the FPGA does not retain any connections or information when it is not powered. The code which governs how it works is stored in external non-volatile memory and is loaded into the FPGA when you turn the radio on. Firmware updates overwrite the information in the non-volatile memory and then the new file is loaded into the FPGA when the chip is reset.

If the random function has been turned on in the ADC, the first thing the FPGA software does is remove the randomisation of the incoming data stream. The next stage is called the CORDIC after the algorithm used to generate the sine and cosine local oscillator signals. The CORDIC performs the functions of a numerically controlled oscillator and mixer.

The input signal to the FPGA is a 16 bit data stream containing the sampled spectrum from 0 Hz to the Nyquist frequency of 61.44 MHZ, (half the 122.88 Msps sample rate). Actually the frequencies above 55 MHz have been filtered out by the low pass filter on the Hermes board and whatever optional low pass filters are engaged on the RF filter board. The low frequency cut off is affected by the input design and the optional high pass 'Alex' filters. The data is coming out of the ADC at a rate of 1.966 Gbits per second. It is not practical to send such a lot of data over a USB or Ethernet connection to the PC, so current SDR models use a mixer coded into the FPGA to select only a part of the HF spectrum. The new FlexRadios get around this problem by performing the DSP functions inside the radio, dramatically reducing the amount of data that needs to be sent to the PC. The FPGA in the ANAN-100 can select up to seven receiver panadapters between 48 kHz and 384 kHz wide. In the following example we will assume that the user has selected a single 192 kHz panadapter. Reducing the bandwidth down to panadapter sized chunks has the advantage of reducing the data rate to the PC. We also get a handy improvement in the receiver's dynamic range and the signal to noise performance. This effect is covered in more detail in the next chapter.

The local oscillator and mixers are created in software. They manipulate the digital data stream mathematically but they work just like hardware oscillators and mixers would. It is a direct conversion process just like the Tayloe mixer in a QSD type SDR. A big problem with QSD receivers is the difficulty in completely cancelling the image signals. For perfect image cancellation the I and Q streams must have the same amplitude and the phase angle between them must be exactly 90 degrees. Even small amplitude or phase errors cause poor image rejection. To compete with the performance of the best conventional receivers we need at least 100 dB of image rejection. If we are to meet the challenge, the amplitudes must be within 0.0001 dB and the phase error must be less than 0.001 of a degree. This is not possible with a hardware solution consisting of a real oscillator and mixers, but it can be achieved using software code in the FPGA.

The CORDIC algorithm calculates sine and cosine functions at the required local oscillator frequency and these are mixed with the incoming signal to produce I and Q signals extending from the local oscillator frequency up. Everything done in the FPGA is achieved by manipulating digital bits using mathematical functions, but it is easier to describe the functional blocks as they would be in an analog design. In fact, it is possible to build a hardware oscillator convert it to a digital signal using an ADC and then inject the output data into the FPGA for use as the local oscillator, but that would introduce another possible noise source. Initial experiments with the Mercury predecessor to the Hermes board used this approach.

After the CORDIC stage the data is still at the 122.88 Msps rate, but it is shifted by the mixer so the output only contains frequencies in and above the panadapter we want. We don't need the frequencies above the wanted panadapter and we don't need such a high data rate to describe a 192 kHz bandwidth so we throw a lot away. This is done in four stages using a process known as decimation and filtering.

To get from a sample rate of 122.88 Msps to a sample rate of 192 ksps, we need a decimation ratio of 122880 / 192 = 640 so we are effectively throwing away 639 of every 640 samples.

We have a problem with Mr Nyquist. The low pass filters before the ADC removed all of the signals above 55 MHz, but the decimation process causes new problems. For example, if we use a decimation ratio of 2 we will halve the sample rate from 122.88 Msps to 61.44 Msps and all of the unwanted signals from, 30.7 MHz to 55 MHz will be aliased back into the new receiver pass band. Once there, they can't be removed so they need to be filtered out before the decimation. This happens every time you decimate the sample, all the way down to our final rate of 192 ksps. Every time we decimate the data rate we need to filter out the new alias frequencies. Finally we want a 192 kHz wide panadapter spectrum display, centred on the local oscillator frequency, with all signals above the 192 kHz bandwidth filtered out.

The developers use CIC (cascaded integrated comb) filters because they are easy and efficient to code in the FPGA. One of the advantages of the CIC filters is that decimation can be built into the same code block as the filter. The first CIC filter is a three stage filter with a decimation ratio of 10. The output is at 12.288 Msps, one tenth of the original rate. This means the filter needs to attenuate all frequencies above 6.144 MHz. Reducing the bandwidth by a factor of 10 gives us 10 dB more dynamic range through process gain, $10 * \log_{10}(122.88/12.288) = 10$ dB. We need more bits to describe the additional dynamic range so the number of bits per sample is increased to 24 bits. This is enough for all further stages so it remains at 24 bits through the rest of the process.

The decimation ratio of the second and third CIC filters can be changed to provide the wanted output bandwidth; 48 kHz, 96 kHz, 192 kHz or 384 kHz. For 192 kHz the second (5 stage) CIC filter would decimate by 8, down to 1.536 Msps and the third (12 stage) CIC filter would decimate by 4, down to 384 ksps. At the end of this block there has been a total of 20 stages of CIC filtering which is enough to reduce aliased signals from above the wanted panadapter by more than 100 dB. One more decimate by two stage will bring us down to the wanted 192 ksps sample rate and bandwidth. But the final stage uses a different kind of filter.

The CIC filters are easy to code and don't use too many FPGA elements but the output pass band is not flat. After the 20 CIC filter stages, the output will have a roll off of about 8 dB at the high frequency end. This is not desired as it would show as a roll off of the noise floor and receiver sensitivity at both edges of the spectrum display. The final filter and decimation by two, down to 192 ksps in our example, is a CFIR (compensating finite impulse response) filter. The FIR filter has a flat response and a sharp roll off above the wanted pass band. A compensating 'CFIR' filter is designed to have a shape which rises at the higher frequencies and then has a sharp roll off. Combined with the CIC filters the output is flat across the wanted receiver panadapter and has at least 100 dB rejection of signals above the cut off frequency.

All of these filters were designed using software like MATLAB and the resulting Verilog code is loaded into the FPGA as a firmware file. They operate mathematically on the data bits representing the signals which have been sampled by the ADC, but their performance can be measured and displayed just like hardware filters. Verilog is a 'hardware description language' (HDL) with a syntax similar to the C programming language.

When the signal is finally converted from strings of numbers back to analog, it has been filtered in exactly the same way as real filters, but without ringing or other artefacts like intermodulation. This is how SDR receivers can create very narrow filters with exceptionally good filter shapes.

That pretty much wraps up what happens inside the receive side of the radio. The bits containing the panadapter data are put into the USB data frames, then combined into EP6 Ethernet frames and sent to the PC. The remaining DSP process culminating in the spectrum display and demodulation of the signal so you can hear all of your mates on the 80m net; is performed by the SDR software on the PC. Of course I have not mentioned anything about the clock, memory, power supply, control and interface chips on the radio board, but we are only trying to understand the theory of operation not build a radio from the ground up.

On the transmit side, the microphone audio is converted to a digital signal and sent to the PC software as a part of the EP6 Ethernet frame, which also includes the data for up to seven receiver panadapters. If an SSB, AM or FM signal is to be transmitted, the microphone audio is processed in the PC software and then sent back to the Hermes board in the EP2 data stream as a modulated signal. If a CW or digital mode signal is intended, the microphone audio is not used and the data in the EP2 packets is CW or digital mode modulation. The signal to be transmitted arrives at the Hermes board as 48 ksps I and Q signals. In the FPGA, the I and Q data streams need to be increased from 48 ksps to 122.88 Msps so that the Nyquist requirement can be met when the RF signal is created. This is achieved using a variant of the CIC filters used in the decimating and filtering process in the receive chain. This time they are Interpolating filters. The Interpolating CIC filter does the reverse of decimation. In this case it creates 2559 new samples in between each input sample. The number of bits per sample is also reduced from 16 bits to 14 bits to match the 14 bit AD9744 DAC (digital to analog converter). The bandwidth of the transmit signal at the output of the Interpolating CIC filter, known as the digital baseband, is about 0.8 x the input sample rate or 38.4 kHz. This, digital baseband, needs to be translated up to the frequency you want to transmit on. The up-conversion uses another CORDIC oscillator and two mixers, the same as the CORDIC section in the receiver part of the FPGA. The I and Q digital baseband signals are mixed up to the wanted transmit frequency and combined into a single 14 bit data stream at the 122.88 Msps data rate. Then the digital signal from the FPGA is converted to an analog RF signal using the AD9744 DAC. From there it passes through a 56 MHz low pass filter to the Hermes power amplifier which boosts the signal up to 500 mW. After the Hermes board the transmit signal is routed to the PA/Filter board where it is amplified to 100 Watts and passed through the Alex low pass filters and on to the antenna connector.

The 38.4 kHz digital baseband is wide enough to carry any conceivable amateur radio transmission. Most modes use a bandwidth less than 3 kHz. The actual bandwidth of your transmitted signal is governed by the modulation method and filtering in the SDR software on the PC.

The Hermes transceiver board is full duplex so both the receiver and transmitter are always active. A transmit / receive relay on the PA/Filter board allows you to use a single antenna for transmit and receive. But if you use two antennas you can use the full duplex mode. This means you can monitor your own transmit signal (suitably attenuated of course). Being a full duplex radio and having a wide digital baseband makes advanced techniques like adaptive transmit pre-distortion possible.

Chapter 6 – Myths about the dynamic range of SDR receivers

There is a lot of rubbish on the Internet about the maximum achievable dynamic range of SDR receivers. I have seen comments suggesting all SDR owners are lying or ignorant about the dynamic range of their SDR equipment. As usual these comments come from people who have little or no experience using SDRs and a poor understanding of how they work.

The numbers in this chapter are for receivers with no antenna connected. When you connect an antenna you will hear and see received noise in addition to the noise created inside the receiver.

One of the dreams of SDR developers has been being able to sample the entire HF spectrum and have all of the signals available to the PC software at once. But two major problems prevented this from being realistic at a reasonable price with the available ADC technology. One problem was the sustained data rate that the DSP section of the receiver needs to handle and the second issue is the limited dynamic range and noise performance of the ADC chips. Both of these issues have now been overcome.

If we assume a sample rate of 122.88 Msps and 16 bits per sample we need to be able to handle the data coming out of the ADC at 1.966 Gbits per second. The FlexRadio, samples at 245.76 Msps so it needs to manage 3.932 Gbits per second. A modern PC can handle the workload but it is not easy transferring data to a PC at such a high sustained rate. FlexRadio gets around the problem by performing the DSP processing on the radio board rather than in the PC. Other amateur band DDC transceivers and receivers sample the entire HF spectrum, but only a part of the spectrum is presented to the DSP process in the PC for demodulation and the spectrum display. The ANAN radios can send up to seven 384 kHz wide panadapters over the Ethernet connection to the PC. A 384 kHz panadapter is wide enough to display the entire 20m band and you normally zoom down to a much smaller area when you are actually operating, so this limitation is no big deal. Using a DSP stage in the radio, or only transferring panadapters to the PC instead of the full HF spectrum, substantially reduces the amount of sustained data required. The inadequate signal to noise ratio and dynamic range of the ADC is compensated by process gain. As we narrow the signal down to the final receiver bandwidth, or the FFT bin size when we are considering the spectrum display, we get an improved signal to noise ratio and more dynamic range.

The dynamic range of a direct down conversion SDR receiver is predominately affected by;

1. The effective number of ADC bits (ENOB) calculated from the published SINAD, not the number of bits the ADC uses to output the data
2. Process gain arising from the ratio of the sampling frequency to the final receive bandwidth, and
3. The number of large signals being received, (actually the sum power of all signals arriving at the input to the ADC)

The noise floor displayed on the spectrum display with no antenna connected is affected by

1. All of the above, plus
2. The bandwidth of the FFT (fast Fourier transformation) bins displayed, and
3. A constant (k) determined by the type of Windowing in use.

As long as the noise floor displayed on the spectrum scope is lower than the noise level within the receiver bandwidth, as indicated by the S meter, you will be able to see any signal you can hear.

The maximum possible sampling dynamic range for an ADC is related to the number of bits used to describe each voltage level in the received signal, DR =20*log(2^Nbits-1) dB. The noise generated by the ADC is predominantly due to, thermal noise in the semiconductors, quantisation errors, and jitter. Some mathematical wizardry produces the formula for the maximum possible signal to noise ratio for an ADC which is also related to the number of bits used, SNR = 6.02*Nbits+1.76 dB.

It is this formula which causes all the aggravation in online forums because on first glance it indicates that the maximum achievable SNR (signal to noise ratio) is not very good and nowhere near the figures quoted in SDR product documentation and reviews. It also shows a significant difference between the performance of the 8 bit, 14 bit and 16 bit ADCs used in various SDR models which is much less apparent in real world tests.

Theoretical ADC performance

bits	DR (dB)	SNR (dB)
8	48.1	49.9
10	60.2	62.0
12	72.2	74.0
14	84.3	86.0
16	96.3	98.1
18	108.4	110.1
20	120.4	122.2
24	144.5	146.2

These are **NOT** the results you see in the real world! A typical 14 bit or 16 bit ADC is likely to have a real SNR of 75 - 77 dB. But adding process gain means that the SDR receiver can achieve real world performance of around 120 dB of dynamic range and SNR in a 2.4 kHz bandwidth.

To match the performance of the best conventional receivers we want a dynamic range in the order of 120 dB. From the table above it looks like this can only be achieved by using fast 24 bit ADCs which are very, very expensive. All the currently available direct down conversion SDRs only use 8, 14 or 16 bit ADCs, so how do they achieve acceptable dynamic range and signal to noise performance?

The answer is 'process gain'. As the bandwidth of the signal is reduced from the Nyquist bandwidth of the ADC (61.44 MHz for the ANAN and 122.88 MHz for the FLEX-6000 series) to the final receiver bandwidth of around 2.4 kHz for an SSB receiver, or the FFT bin width of less than 100 Hz for the panadapter spectrum display, we get a large improvement in the dynamic range and hence the signal to noise ratio of the signals we are receiving.

The major market for fast ADC chips is for military and cellular base station radios using digital modulation methods, where the signal to noise performance and dynamic range are not as critical as they are for amateur radio. Designing fast 24 bit low noise ADCs is not a high priority for the manufacturers. New ADCs are likely to have faster sampling rates and a wider bandwidth, rather than more bits.

The current DDC type SDRs use a local oscillator and mixers coded into the FPGA to select parts of the HF spectrum. The FlexRadios can select up to eight panadapters 14MHz wide. The Apache radios can select up to seven panadapters 384 kHz wide. Reducing the bandwidth to 14 MHz or 384 kHz increases the dynamic range of the signals and allows the data sample rate to be reduced. Slower data is easier to transfer to the PC and easier for the DSP section to process. Multiple panadapters can be processed in parallel allowing the SDR software to demodulate and display signals on several panadapters at the same time.

The noise component in the ADC samples is spread pretty evenly across the whole of the sampled bandwidth. Reducing the bandwidth from 61.44 MHz down to 384 kHz improves the signal to noise ratio in the same way as reducing the bandwidth on a spectrum analyser. Another way to look at how this works is to consider that every 16 bit sample includes a small error due to noise. If you throw away many of the samples during the decimation process you throw away the noise associated with them. Either way, reducing the bandwidth reduces the noise floor resulting in better signal to noise ratio and an increased dynamic range, since that is the range from full amplitude to the noise floor. The amount of process gain is directly proportional to the reduction in bandwidth, process gain = 10 * log(Fs/BW). A reduction in the sample rate from 122.88 Msps to 384 ksps, results in a handy 25 dB of improvement in the signal to noise ratio and dynamic range. If we assume, from the table, that the signal to noise performance of a 16 bit ADC is 98 dB then;

98 dB SNR + 25 dB process gain = a new SNR of 123 dB.

*If you have been reading other papers you may have noticed a discrepancy here. Usually the process gain formula uses the Nyquist bandwidth and not the full Fs sample rate. It is shown as either 10 * log((Fs/2)/BW) or 10 * log(Fs/2BW). This formula would be correct except that the SDR uses I and Q streams allowing us to use frequencies both above and below the LO frequency which effectively doubles the bandwidth. This is the same in QSD SDRs which display a bandwidth equal to the sample rate, not equal to ½ the sample rate. So technically our process gain = 10 * log((61440+61440) / 384) = 25 dB.*

Because we now have an improved dynamic range we need more discrete levels so the number of bits representing each sample is increased from 16 to 24. This does not present a data handling issue because the sample rate has decreased to 384 ksps. So now we only need a transfer rate of 9.2 Mbits per second for each receiver panadapter.

When we listen to the SSB signal in a 2.4 kHz bandwidth or a CW signal in a 500 Hz or less bandwidth we get even more reduction in noise and an even better signal to noise.

But process gain is not the end of the story because real ADCs do not achieve the maximum possible signal to noise ratio shown on the table. The ADC data sheets quote the real signal to noise in several ways and at several frequencies. They usually quote SNR (signal to noise ratio), SINAD (signal to noise plus distortion ratio), SFDR (spurious free dynamic range) and THD (total harmonic distortion).

The real SNR of the ADC is less than the maximum theoretical SNR so at very low input levels, represented by low binary numbers, the ADC is unable to accurately represent the input voltage.

A very small input signal where only bits 0, 1 and 2 of the output data are not zero will be masked by the noise generated inside the ADC and may not be sampled accurately. In other words the effective number of bits (ENOB) is less than the actual number of bits describing the output data. This is very important because it means the SNR calculation must be done using the ENOB rather than the total number of bits.

ENOB can be calculated from the published SINAD figure, ENOB = (SINAD-1.76)/6.02.

The ANAN/Hermes board uses a 16 bit LTC2208 ADC with a SINAD of 77.4 dB giving an ENOB of around 12.6 bits. The Perseus receiver uses a 14 bit LT2206 ADC with a SINAD of 77.1 dB giving an ENOB of around 12.5 bits and the AD9467 ADC used in the new FlexRadio SDRs has a published SINAD of 76.2 dB and an ENOB of 12.4 bits. This means even though the Perseus uses a 14 bit ADC the actual dynamic range performance is almost the same as the 16 bit ADC used in the ANAN radio. The main difference is in the sample speed which allows the ANAN to work to 55 MHz while the Perseus only works to around 40 MHz.

To wrap things up, you can't really judge a direct down conversion SDR based on the number of bits its ADC uses. The key is to choose a radio that uses a very low noise ADC with a very high sample rate. Because the dynamic range and noise level in the receiver bandwidth you are listening to is dependent on the SINAD published on the ADC data sheet and the sampling rate, not the actual number of bits in the output data. Since the SINAD determines the effective number of bits, you can ignore the SNR calculation and table at the beginning of the chapter. The sample rate determines the maximum 'Nyquist' bandwidth for the receiver and the ratio between the sample rate and the final receiver bandwidth determines the process gain.

The noise level in bandwidth BW = -(SINAD + 10 * log(Fs/BW)+1).

Noise level in 2.4 kHz BW = -(77.4 + 10 * log(122880/2.4)+1) = -125.5 dBm [ANAN S meter -125.8]

Noise level in 500 Hz BW = -(77.4 + 10 * log(122880/0.5)+1) = -132.2 dBm [ANAN S meter -133.2]

This calculation gives you the noise level expected in the final receiver bandwidth with no antenna connected to the receiver. It seems to relate well to the S meter reading of the ANAN receiver on the 20m band with no preamp and no dither. Adding dither will raise the noise level a dB or so but improves the IMD performance. Using the random function reduces spurious responses without much impact on the noise level. The FlexRadio has double the initial sample rate so you would expect a 3 dB improvement in the final noise level.

The +1 dB in the above formula is because the SINAD is quoted using a signal 1 dB below full scale. If you don't know the SINAD you can use SINAD = 6.02 estimated ENOB +1.76.*

Note: In this chapter I have tried to write the formulas so they can be easily pasted into a spreadsheet. This is why the multiply symbol is a * not an x or omitted. The caret ^ symbol indicates an exponent.

ADC gain compression

What happens when you connect the antenna? Well, the noise level shown on the spectrum display increases, for two reasons. Firstly the antenna picks up all kinds of; atmospheric noise, electrical noise, splatter and harmonics from transmitters, and in my case interference from my ADSL modem. Secondly, the dynamic range of the ADC is affected by the total power of all the signals presented to it. Receiving a lot of large signals reduces the dynamic range and the displayed noise floor rises. We don't want any signals from above the Nyquist frequency and we also don't want unwanted signals, such as AM broadcast stations, reducing our receiver's dynamic range and signal to noise ratio. Having a few large signals, from within the Nyquist bandwidth will not affect the noise floor by too much and will act to reduce the effects of intermodulation distortion.

In many SDRs, hardware filters are installed before the ADC to protect the receiver from gain compression and to combat possible out of band or in band interference. They must be designed and made with care so they do not add any new noise or cause intermodulation problems. The ANAN design includes a 55 MHz low pass filter on the Hermes board, plus the 'Alex' high and low pass filters unless they are bypassed. The low pass filters on the PA/Filter board are primarily for the transmitter, but they are also used for receiving. The high pass filters are used for receiving only and they can be switched out if you want to make the receiver wide band. RX2 on an ANAN transceiver has only the 55 MHz LPF.

The spectrum display

The spectrum display is much wider than the bandwidth of the audio channel you listen to, which is typically around 2.4 to 3 kHz for an SSB signal and 200 to 500 Hz for a CW signal. But the noise level displayed is lower than the noise level in the audio signal and as indicated on the S meter. A wider signal bandwidth has less process gain so the displayed noise floor should be higher, but it is not. How can that be? What magic is happening? The answer is the use of a technique initially developed for spectrum analysers.

A spectrum analyser displays the level of signals across a range of frequencies and the panadapter spectrum display on an SDR does the same thing. Older spectrum analysers display the signal level received by a narrow band receiver being continually tuned across a wide frequency range. The persistence of the display allows you to see the spectrum display across the 'swept' band. In newer spectrum analysers the narrow bandwidth is converted to a digital signal and stored in memory so the whole display can be viewed at once and updated more rapidly. This results in a much brighter display with less flicker. The latest spectrum analysers and the display on your SDR receiver use a process called fast Fourier transformation (FFT) which converts the sampled data from a time domain signal as would be viewed on an oscilloscope to a frequency domain signal as viewed on a spectrum analyser. Rather than converting one narrow bandwidth signal into a digital signal and storing it, FFT divides the digital information representing the spectrum, in our case a panadapter, into thousands of very narrow 'bins'. The contents of each bin are stored and displayed at a high rate on the spectrum display. Because the bins are only a few Hertz wide, the noise level in each bin is low. The display is made up of many bins side by side, so the overall displayed noise floor is also low.

The actual noise floor displayed with no antenna connected to the receiver is not important provided it is lower than the noise level on the panadapter and in the bandwidth you are listening to when the antenna is connected. As long as it is, you can be sure you will be able to see any signal on the panadapter that is strong enough to be heard.

The displayed noise floor with no antenna connected is affected by; the sampling rate of the 48 – 384 kHz panadapter, the number of bins this is divided into, which sets the width in Hz of each bin, the constant associated with the windowing filter and of course the actual noise performance of the ADC and FPGA.

I have looked for a formula or explanation of how to predict the panadapter noise floor but I have been unable to find any useful explanation of how this works in the ANAN radio / PowerSDR mRX combo.

Changing the windowing type has a small effect on the noise floor, around a dB or so and halving the sample rate improves the noise floor by 3 dB as predicted by the process gain calculation. Most of the available information online indicates that halving the bin width should also improve the noise floor by 3 dB, on the basis that something half as wide should let in half the amount of noise, but this does not happen. I only get an improvement of about 1.8 dB each step. This may be due to the shape of the windowing filter or other processes. Even after extensive online research, the reason for this discrepancy remains a mystery to me.

ADC Noise

The noise generated by the ADC is predominantly due to; thermal noise in the semiconductors, quantisation errors and jitter. At each sample time the ADC measures the voltage of the input signal. It decides that the sampled voltage lies somewhere between two nominated voltage levels and outputs a 16 bit number representing the relevant voltage step. The difference between one voltage step and the next is very small but there is always a small error between the quantised (sample) level and the real input level. They are known as quantisation errors because they are errors in quantifying the exact input voltage. Quantisation errors cannot be corrected. Since the errors are random, the difference between the input signal and the sampled signal is called quantisation noise. The only way to reduce the quantisation errors to zero would be to use an infinite number of individual voltage levels which would require an infinite number of bits. Jitter causes random phase errors in the timing of the sampling but this is minimised by using a very clean 'low noise' clock signal. Thermal noise is insignificant compared to quantisation noise. ADCs using more bits to describe each voltage level can represent more individual voltages and therefore have less quantisation noise.

Dither and random

Dither and Random also relate to the ADC. They are selectable in PowerSDR mRX. Whether you should have them turned on depends on how you use your radio and whether you have high power stations near your QTH. I don't live in a particularly noisy region and I don't notice much difference with either function on or off, so I usually leave both functions turned off. My explanation on how dither and random works is included in chapter 5.

Phil Harmon explains the function of dither and random very well in his videos, I urge you to watch them. The videos are available on YouTube as Ham Radio Now episode 62 parts 1, 2 and 3. They are great viewing on a wet winter day.

To recap briefly; dither improves the intermodulation performance of the ADC by ensuring the input signals are well spread across the analog to digital transfer characteristic of the ADC. Noise is added to the analog signal and then removed after the analog to digital conversion. In the real world with the antenna connected there are many signals being processed by the ADC so using the dither function is probably not necessary.

Random is used to remove spurs on the panadapter which are caused by the ADC sampling the same pattern of signals over and over. Unintentional coupling of the output data and the clock signals into the sensitive analog input can cause spurs to build up on the output signal. The already random nature of signals on the ham bands ensures that the digital output is random and no spurs will be created. The random function will remove spurs which might become visible when looking at a quiet band or on shortwave bands, where the signals are continuous. Many ADCs do not include the internal application of dither and random. The functions can be added with other hardware or not used at all.

Is there process gain when using a QSD type SDR receiver?

A QSD type SDR relies on a PC sound card, or an audio frequency ADC, for the analog to digital conversion. Because the sample rate is much lower, there is not as much process gain when the signal is reduced to the 2.4 kHz pass band of the receiver, or the FFT bin width of the panadapter spectrum display. So the dynamic range and SNR improvement is much less. For example, the process gain 10 * log(Fs/BW) for a 48 kHz sound card is 10 * log(48/2.4) = 13 dB. For a 96 kHz sound card it rises to 16 dB. It is quite low compared to 10 * log(122880/2.4) = 47 dB of process gain for the ANAN DDC receiver.

In summary;

1. The internal SINAD (signal to noise + distortion) performance of an ADC is more important than the number of bits its output uses
2. The SNR = 6.02*Nbits+1.76 formula only tells a part of the story. Process gain improves the signal to noise ratio and dynamic range to levels which can match the best conventional receivers
3. The spectrum display is wider than the demodulated bandwidth but it can display a low noise floor due to the magic of FFT
4. Random and dither may help, but may be more of an issue during lab tests than on active ham bands
5. Receiving medium size signals can actually improve the receiver's intermodulation performance and make it easier to hear a weak signal, but
6. Receiving a lot of very large signals may compress the dynamic range
7. Good quality front end filters before the ADC protect the dynamic range and prevent break through from aliased signals above the Nyquist bandwidth.

Chapter 7 – SDR software on the PC

- Section 1: Introduction, frame structure, key software modules and FFT
- Section 2: The wideband spectrum process
- Section 3: The transmit process
- Section 4: The receive process
- Section 5: Command and control
- Diagrams:
  - Ethernet frame structure
  - Reversal of the FFT output buffer
  - Kiss Konsole functional block diagram
  - EP6 USB frame structure
  - EP2 USB frame structure
  - EP4 wideband spectrum data
  - Command & Control bytes from the radio
  - Command & Control bytes to the radio

Introduction to the PC software, frame structure, key software modules and FFT

The PC software, whether it is running on a Windows, Linux, Macintosh or Android based device, is a huge part of almost all SDR receivers and transceivers. The exception is SDRs 'with knobs' which don't need a PC and can be operated like conventional architecture radios.

In most cases the SDR software manages the control of the radio functions, provides the operating panel, creates the spectrum and waterfall displays, and carries out the modulation, demodulation and other DSP functions like noise and notch filtering. It is the last piece in the SDR puzzle.

So far I have covered the fundamentals of SDR architecture, the expected performance of SDR technology and what happens inside a typical DDC transceiver such as the Hermes board. I have a strong desire to understand these new radios and why they work as well as they do. I decided after spending a lot of time and effort learning about software defined radio that I should attempt to pass the information on to other interested hams, so this book is a by-product of my learning experience. Hopefully it will make learning about SDR easier for you.

To research the information covered in the first six chapters, I used a variety of Internet sources and also Phil Harmon's excellent series of videos about the Mercury receiver. But my online search hit a wall when I tried to find out how the PC software works. This is rather odd because there are many SDR software applications available and most of them are open source, so I expected there to be a lot of reference information. There seems to be a gap between the skilled software developers and the general ham population who are for the most part happy to use the software without understanding how it works inside. I seem to fall into a small group (of one?) who wants to know how it works, but who does not want to create their own.

One of the nice things about using software defined radios is the ability to use different PC software with the same SDR hardware. There are dozens of SDR software packages but I couldn't find any documentation in the form of tutorials, flow charts or functional block diagrams. This is not a criticism, I guess the software guys are more interested in writing the applications and they find it easy to read the code so they don't need flow charts and diagrams.

There is a trend towards developing server / client SDR packages which allow a 'server' part of the code to reside on the PC connected to the radio while the GUI (graphical user interface) 'client' can reside on any PC in the local network. This means you can operate your radio from a wireless device or maybe even a phone. FlexRadio has taken this idea to the point where the server and DSP functions are inside the radio and the GUI application resides on any PC connected to the local network. You can even run multiple instances. So for example, you can watch for a 6m band opening on your phone, while working 40m PSK-31 on your laptop, with the computer in the shack running CW skimmer on 20m. Ultimately you will be able to use your SDR remotely from any computer or tablet device attached to the Internet.

To help me understand a typical SDR PC application, I looked at GNURadio which allows you to build SDR software using blocks of code developed by yourself or other contributors. This is very cool, but is primarily Linux based and not really aimed at my OpenHPSDR, Hermes based radio. I also looked at an old version of PowerSDR written in Visual BASIC, but I abandoned that as well because it was out of date. Finally I turned to Kiss Konsole which was developed in C# (C Sharp) specifically as an open source tutorial and test platform for people wanting to experiment with writing their own SDR application. As a result it is fairly well commented. I have never used the C or C# programming languages although I have written several applications using Visual BASIC, so I have had quite a struggle teasing out how Kiss Konsole works.

Kiss Konsole is not a server / client based application and it is not as fully featured as the latest software such as PowerSDR mRX, Studio 1, or cuSDR. But it does do the fundamentals and that is all we need to get an appreciation of what is done by the PC half of your SDR radio. Different SDR software will have different code blocks and features but the general scope will be similar. The radio in this analysis is my trusty Apache ANAN-100 which uses a Hermes SDR transceiver board. The information provided will also be relevant to stand-alone Hermes boards, the Angelia board used in the ANAN-100D and the new Orion board used in the ANAN-200D. A lot of the clever DSP stuff in Kiss Konsole is done by SharpDSP which was written and kindly made available for free by Phil Covington one of the pioneers of SDR for hams. The Kiss Konsole code is a team effort. There is an acknowledgement of the copyright holders for both SharpDSP and for Kiss Konsole at the start of the book and on the functional block diagram.

Now would be a good time to have a look at the block diagram on pages 44 and 45, so you can follow along. The grey boxes are Kiss Konsole code and the other colours are SharpDSP code. The FFTW.dll box is the Fast Fourier Transformation code used by SharpDSP. I believe it is an application called, "Fastest Fourier Transformation in the West" (FFTW). A dll file contains reusable software subroutines that can be used by other programs. This allows software developers to avoid reinventing the wheel. Under licence conditions they can incorporate code that another developer has written, to into in their software applications.

The **graphical user interface (GUI) - Form1.cs** is responsible for how the program looks and works. It is the knobs and front panel of the transceiver. The GUI has a 'screen' to display the panadapter which can be 48 kHz, 96 kHz, 192 kHz or 384 kHz wide on the band you have selected, and either the wideband spectrum up to around 55 MHz, or a waterfall display. There are many check boxes and slider controls for setting all the usual controls; volume, squelch, AGC, frequency etc. and there is an S meter. On the menu there are several setup screens for less used functions. Form1 also reads and writes to the KKCSV text file which stores program settings so when you change bands or turn the software off, the radio remembers its previous settings. This is called persistence. Form1 talks mostly to HPSDRDevice.cs which controls the flow of signals through the application. It also gets and sets information to and from the SharpDSP process by talking primarily to the Receiver.cs code block. A large part of the code in Form1 is dedicated to drawing and updating the panadapter and the waterfall or wideband spectrum display.

Ethernet.cs manages the Ethernet data to and from the radio. In earlier OpenHPSDR equipment the interface to the radio was via a USB port, but the interface to the Hermes board used in Apache radios is over an Ethernet connection, the same as the network connection to your Internet router. There are three types of data streams or frame structures used to communicate with the Hermes board.

- All of the Ethernet frames begin with two synchronisation 'sync' bytes 'EF FE', one status byte which is always 01, one EP byte set to 02, 04 or 06, and a 32 bit (4 byte) frame number.
- EP2 frames carry data from the PC to the radio. In Kiss Konsole each EP2 frame includes two 512 byte USB frames. The data in the USB frame includes; control and command information to the radio, left and right audio for the speaker connection and the I and Q data streams used for the transmitter. The software has the option of sending four USB frames but this is not implemented, so each 1032 byte EP2 Ethernet frame carries 1024 bytes of payload data.

Sync EF FE (2 bytes)	Status 1 (1 byte)	EP 2 (1 byte)	Sequence Frame No. (4 bytes)	EP2 USB Frame (512 bytes)	EP2 USB Frame (512 bytes)

- EP4 frames carry data for the wideband spectrum display from the radio to the PC. The 1024 byte data payload is 512 x 16 bit samples direct from the ADC, (not I and Q data).

Sync EF FE (2 bytes)	Status 1 (1 byte)	EP 4 (1 byte)	Sequence Frame No. (4 bytes)	EP4 512 x 16bit ADC samples (1024 bytes)

- EP6 frames carry data from the radio to the PC. The EP6 frame includes four 512 byte USB frames. The USB frames carry control and command information from the radio, 16 bit audio samples from the microphone plugged into the front panel of the radio, and I and Q data for up to seven panadapters. Each 2056 byte EP6 Ethernet frame carries 2048 bytes of payload data. Panadapters are called receivers in Kiss Konsole.

Sync EF FE (2 bytes)	Status 1 (1 byte)	EP 6 (1 byte)	Sequence Frame No (4 bytes)	EP6 USB Frame (512 bytes)	EP6 USB Frame (512 bytes)	EP6 USB Frame (512 bytes)	EP6 USB Frame (512 bytes)

Phil Harmon VK6APH is currently working on streamlining the Ethernet frame structure to make it more efficient, so the PC software and Hermes firmware will be updated to accommodate the changes.

Ethernet.cs runs as a separate thread independently of the rest of the software. Every time an EP2 buffer is full, it sends an EP2 frame to the radio and every time an EP4 or an EP6 frame is received it is stored in a buffer which is read later by the HPSDRDevice.cs block.

If the wideband spectrum has been requested, the radio (Hermes board) will send a block of either 8 or 32 EP4 frames containing the wideband spectrum samples. Each EP4 Ethernet frame contains 512 x 16 bit samples (1024 bytes). When the 8 kbytes or 32 kbytes of data has been stored into the *EP4buf* buffer, the buffer full variable 'flag' is set and the HPSDRDevice block will be able to read the buffer and eventually send it to the wideband spectrum display.

When EP6 frames are received, four 512k USB frames (2048 bytes) are stored into the *rbuf* buffer. When the buffer is full the buffer full variable 'flag' is set and the HPSDRDevice block will be able to read the buffer and process the information.

HPSDRDevice.cs is the gate keeper. It manages the data streams between the SharpDSP processes and the Ethernet block. It sends the spectrum display data to Form1 for display and it also takes control and commands from Form1 to be sent the radio and sends commands and information from the radio to Form1.

HPSDRDevice has two main routines; one called *Process_Data* which handles the EP6 data from Hermes and the other called *Data_send_core* which creates the 512 kB EP2 USB frames to be sent to Hermes. The EP4 wideband spectrum data is routed directly through HSPDRDevice to a SharpDSP process in the OutbandPowerSpectrumSignal.cs block.

The USB frame structure was originally used for the USB interface and is retained for backward compatibility. With the Ethernet interface, the USB frames are combined into the EP2 or EP6 Ethernet frames. The diagrams of the USB frame structure are on pages 46-49.

EP2 USB FRAME STRUCTURE

- The data sent to Hermes in each EP2 Ethernet packet is in two 512 kB USB frames.
- Each USB frame has 3 sync bytes, 5 command and control bytes, and 504 payload bytes made up of 16 bit left and right speaker audio, and 16 bit I and Q transmit audio data.
- <Sync> <Sync> <Sync> <C0> <C1> <C2> <C3> <C4> <Left1> <Left0> <Right1> <Right0> < I1 > < I0 > < Q1 > < Q0 > <Left1> <Left0> <Right1> <Right0> < I1 > < I0 > < Q1 > < Q0 > etc.

EP4 DATA STRUCTURE

- The EP4 data received from Hermes via EP4 Ethernet packets is 512 x 16 bit data samples. Either 8 kB or 32 kB of data is grouped into the *EP4buf* buffer which is read by OutbandPowerSpectrum Signal.
- There is no USB frame structure for EP4 just raw 16 bit samples with no sync or control data.
- < D1 > < D0 > < D1 > < D0 > < D1 > < D0 > < D1 > < D0 > < D1 > < D0 > < D1 > < D0 > etc.

EP6 USB FRAME STRUCTURE

- The EP6 data is in the *rBuf* buffer received from Hermes in each EP6 Ethernet packet as four 512 kB USB frames.
- Each USB frame has 3 sync bytes, 5 command and control bytes, and 504 payload bytes made up of 24 bit I and Q data from each active panadapter and then 16 bits of microphone audio.
- Higher sample rates result in a higher rate of packets being sent from the radio to the PC.
- There can be between one and seven panadapters, which are called receivers in Kiss Konsole. Running more receivers also results in a higher rate of packets being sent from the radio to the PC.
- Where the resultant number of bits does not round out to 504 bytes the frame is padded with zeros at the end of the frame.
- <Sync> <Sync> <Sync> <C0> <C1> <C2> <C3> <C4> <I2^{RX1}> <I1^{RX1}> <I0^{RX1}> <Q2^{RX1}> <Q1^{RX1}> <Q0^{RX1}> <Mic1> <Mic0> <I2 ^{RX1}> <I1^{RX1}> <I0^{RX1}> <Q2^{RX1}> <Q1^{RX1}> <Q0^{RX1}> etc.
- This example assumes one receiver (panadapter), as used in Kiss Konsole. See the drawings later in the section for an example showing the USB frame for three receivers.

Receiver.cs is the main interface between the Kiss Konsole application and the SharpDSP processes written by Phil Covington. It manages the calls to other SharpDSP processes and responds to controls and requests for information from Kiss Konsole. Within the requirements of the GNU General Public Licence, SharpDSP can be used as the DSP core for other SDR application programs. SharpDSP calls on the FFTW.dll for FFT calculations.

What is FFT?

Fast Fourier Transformation (FFT) is an essential resource in any SDR application, but what does it do? It converts signals in the 'time domain' into signals in the 'frequency domain' or back the other way… "Er what the heck does that mean?"

An analog to digital converter samples the level of the incoming wideband received signal at instants in time. Each sample is represented by a 16 bit digital number at the ADC output. Between the ADC and the FPGA, the numbers become a stream of 16 bit data words. Inside the FPGA some decimation and filtering is performed using mathematical wizardry. The result is 24 bit I and Q signals, which are sent to the hardware DSP stage or a PC application. The ADC is very fast. For example the ANAN ADC reads, quantifies, and outputs 122,880,000 samples every second.

The key word here is 'TIME'. The 16 or 24 bit samples represent the received signal voltage at different times. If you converted the signal back to an analog waveform you could plot it on an oscilloscope with voltage on the Y axis and time on the X axis. An oscilloscope shows signals in the 'time domain.'

But we don't want to display the signals that way. We want to see a spectrum display with signals represented at their operating frequencies across the band. The panadapter spectrum and waterfall displays show voltage on the Y axis and frequency on the X axis, essentially the same as the display on a spectrum analyser. A spectrum display shows signals in the 'frequency domain'.

It is often easier to perform functions like filtering on the signals after they have been converted to the frequency domain, so some of the DSP filtering is done after calling the FFT process.

Instead of a series of numbers representing the signal level at different times, FFT creates a series of numbers representing the signal level at different frequencies.

After the FFT process, each sample (numeric value) represents the signal level in a narrow section of the frequency spectrum called a bin. To create something like a low pass filter you can now progressively reduce the level of the bins above the cut off frequency. Making filters becomes easy. By altering the level in individual bins, you can make fine adjustments to the shape, or the high and low cut off frequencies of a filter. Frequency translation can be done simply by moving the position of bins in the buffer.

If we look up the Signal to Noise ratio of the ADC and add the process gain achieved by reducing the bandwidth from 61.44 MHz (Fs/2) down to the SSB bandwidth of 2.4 kHz, we can work out the expected noise floor level of the receiver with no antenna connected. The bandwidth of the panadapter spectrum is somewhere between 48 kHz and 384 kHz which is a lot wider than the 2.4 kHz SSB bandwidth, so you would expect a higher noise level on the panadapter display because there is less process gain. Or put simplistically, think of the difference between a bucket and a bottle left out in the rain. The bucket is wider so more noise (rain) gets in. But we find that the panadapter shows a much lower noise level (with no antenna connected), than the SSB noise level shown on the S meter. This benefit is due to the FFT bins. The panadapter display is effectively showing thousands of very narrow receivers side by side. Each bin has a very narrow bandwidth, usually less than 100 Hz, and therefore each bin has a low noise floor.

In PowerSDR mRX the number of bins used by the FFT and therefore the bandwidth of each bin in Hertz is adjustable, which has an impact on the sharpness and resolution of the spectrum display, especially when zoomed. In Kiss Konsole the number of bins is fixed. The FFT for the Kiss Konsole wideband spectrum display creates 4096 bins and the FFT for the panadapter spectrum display creates 2048 bins. The bandwidth of the bins is equal to the sample rate divided by the number of bins. For example the FFT bin width for a 192 kHz panadapter is, 192000/2048 = 93.75 Hz per bin.

FFT uses very complex mathematics including so called 'real' and 'imaginary' numbers. The maths is outside the scope of this book and it is even too complex for some of the SDR software developers. Luckily there are open source FFT algorithms available in already published dll files. The SDR software developer can write an application that makes calls to the FFT dll file and gets the converted data back in a data array or buffer. You send the FFT routine time related data and it returns frequency related data, or you can send frequency related data and it returns time related data. I believe that most amateur radio SDR software uses an FFT program called, "Fastest Fourier Transformation in the West" (FFTW).

It is very important to understand how the FFT outputs the frequency data and it confused me for quite a while. I had to read some of the "mind bending" FFTW manual before I understood what happens. I am still not sure why the buffer data is arranged the way that it is but I guess that doesn't matter.

The FFT output buffer contains bins above and below the nominal centre frequency. Frequencies above the centre frequency i.e. the frequency the radio is tuned to, are called "real" or "positive" frequencies and frequencies below the nominal centre frequency are called "imaginary" or "negative" frequencies. In Kiss Konsole, the 1st half of the output buffer contains "negative" frequencies starting at 0 Hz and decreasing down to –Fs/2. The 2nd half is "positive" frequencies starting at +Fs/2 and decreasing down to 0 Hz. If the sample rate is set to 192 ksps, the FFT output has 0 to -96 kHz in the first half of the buffer and +96 kHz to 0 Hz in the second half of the buffer. Having the data in the wrong order is clearly no good for a spectrum display. So after the FFT has been run, each half of the buffer has to be reversed in order to get a plus and minus 96 kHz spectrum panadapter display centred on the nominal receiver frequency.

The output of the FFT has to be rearranged for the spectrum display

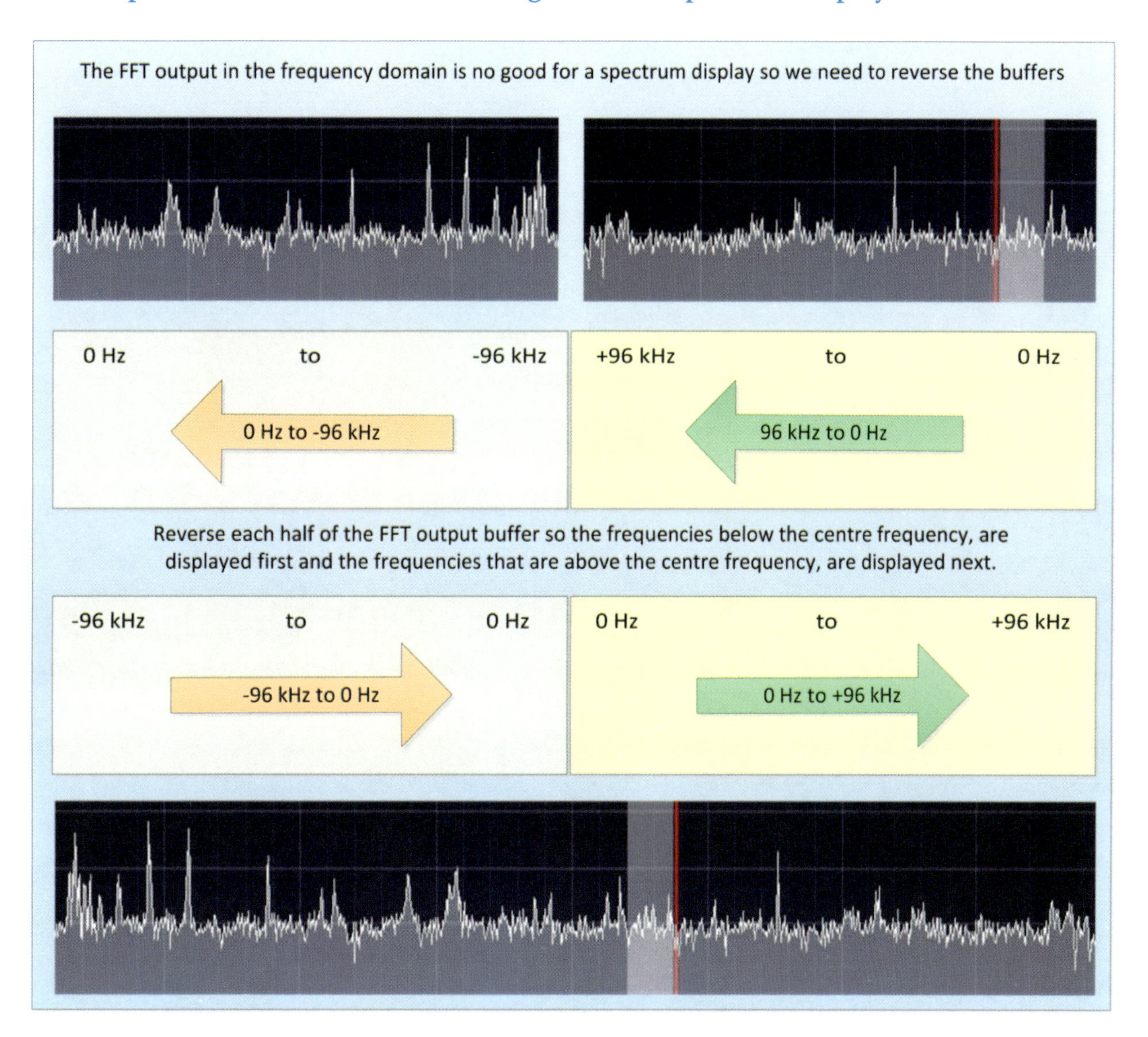

The FFTW manual and other references state that the FFT outputs the positive frequencies from 0 to Fs/2 (0-96 KHz) first and then the negative frequencies from –Fs/2 to 0 (-96-0 kHz). I think that the data is fed into the output buffer on a FIFO (first in first out) basis. So that when the buffer is full it has 0 Hz at byte 0, -96 kHz at byte 1023, 96 kHz at byte 1024 and 0 Hz at byte 2047.

KISS Konsole and SharpDSP Software Modules

Simplified Functional Block Diagram.

By Andrew Barron ZL3DW

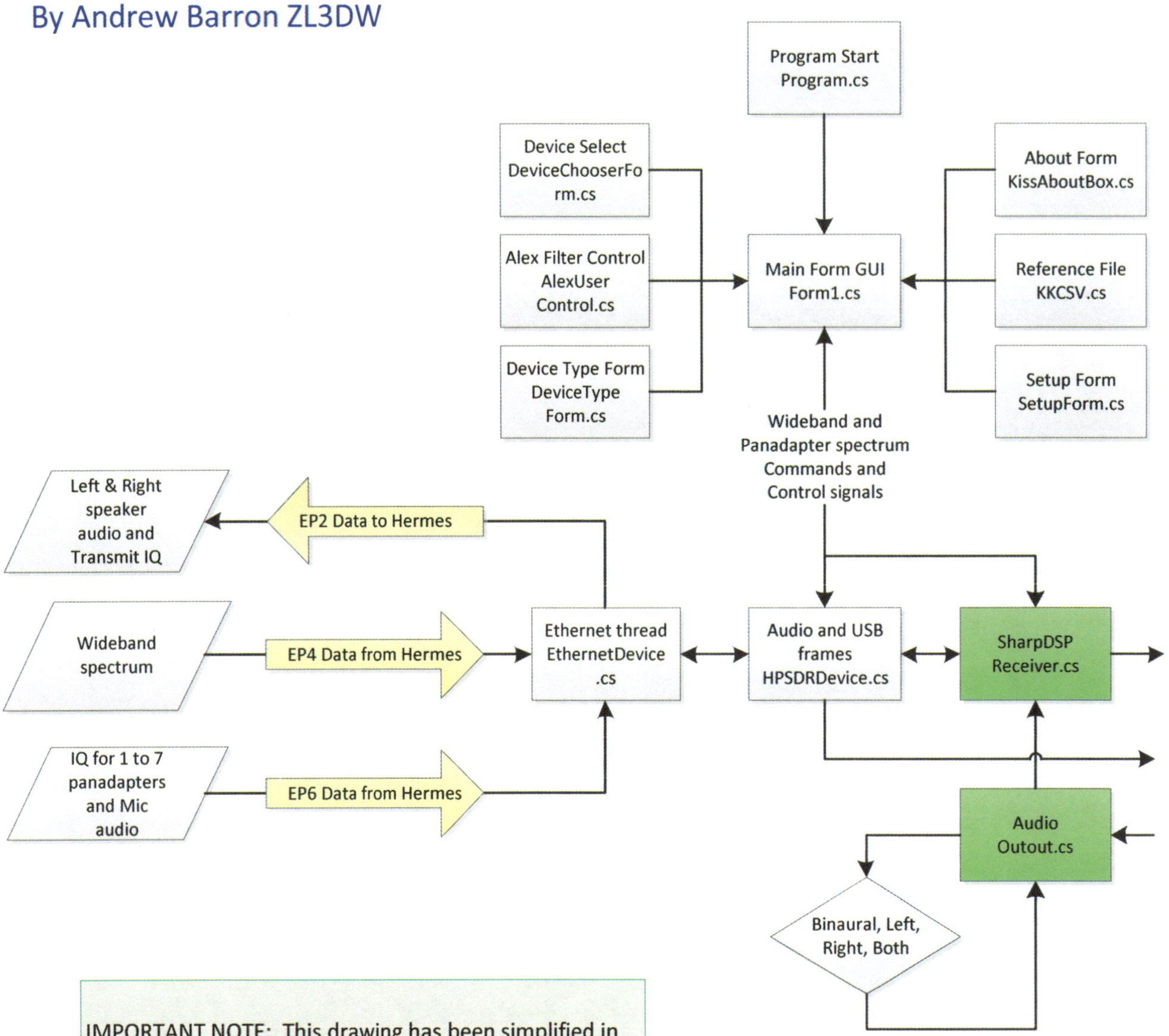

Program Start
Program.cs

Device Select
DeviceChooserForm.cs

Alex Filter Control
AlexUser
Control.cs

Device Type Form
DeviceType
Form.cs

Main Form GUI
Form1.cs

About Form
KissAboutBox.cs

Reference File
KKCSV.cs

Setup Form
SetupForm.cs

Wideband and
Panadapter spectrum
Commands and
Control signals

Left & Right
speaker
audio and
Transmit IQ

EP2 Data to Hermes

Wideband
spectrum

EP4 Data from Hermes

IQ for 1 to 7
panadapters
and Mic
audio

EP6 Data from Hermes

Ethernet thread
EthernetDevice
.cs

Audio and USB
frames
HPSDRDevice.cs

SharpDSP
Receiver.cs

Audio
Outout.cs

Binaural, Left,
Right, Both

IMPORTANT NOTE: This drawing has been simplified in order to illustrate the main functionality of Kiss Console and SharpDSP. There are a few other modules distributed in the package that are unused in a Hermes configuration or of minor importance and there are other connections between modules. The author of this drawing is not a programmer and has had no involvement with the development of this computer code. The author of this drawing accepts no liability in relation to the drawing and does not guarantee the accuracy of the information presented, it is intended as a guide for entertainment only, E&OE.

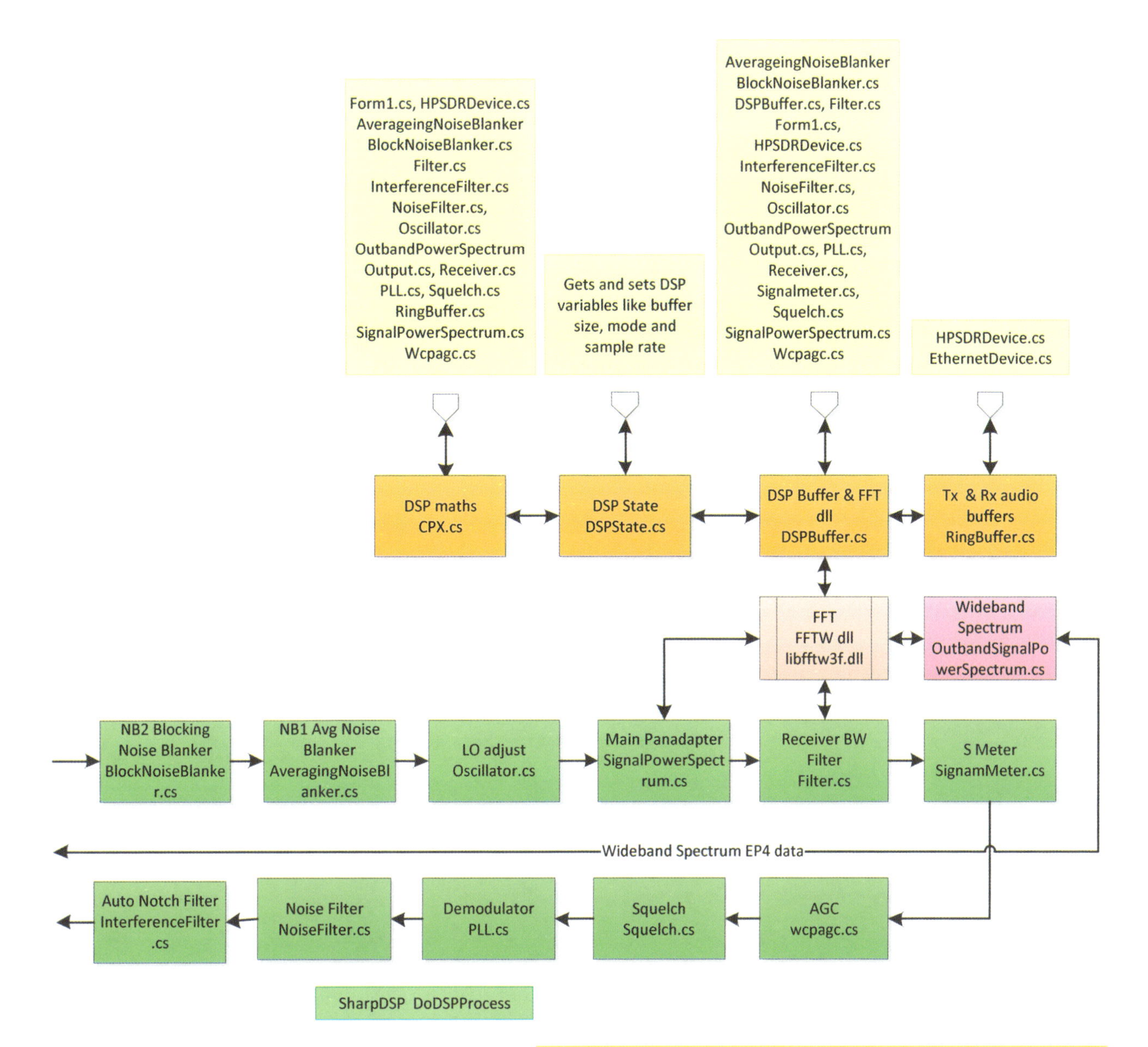

Form1.cs, HPSDRDevice.cs
AverageingNoiseBlanker
BlockNoiseBlanker.cs
Filter.cs
InterferenceFilter.cs
NoiseFilter.cs,
Oscillator.cs
OutbandPowerSpectrum
Output.cs, Receiver.cs
PLL.cs, Squelch.cs
RingBuffer.cs
SignalPowerSpectrum.cs
Wcpagc.cs
Gets and sets DSP variables like buffer size, mode and sample rate
AverageingNoiseBlanker
BlockNoiseBlanker.cs
DSPBuffer.cs, Filter.cs
Form1.cs,
HPSDRDevice.cs
InterferenceFilter.cs
NoiseFilter.cs,
Oscillator.cs
OutbandPowerSpectrum
Output.cs, PLL.cs,
Receiver.cs,
Signalmeter.cs,
Squelch.cs
SignalPowerSpectrum.cs
Wcpagc.cs
HPSDRDevice.cs
EthernetDevice.cs
DSP maths
CPX.cs
DSP State
DSPState.cs
DSP Buffer & FFT
dll
DSPBuffer.cs
Tx & Rx audio
buffers
RingBuffer.cs
FFT
FFTW dll
libfftw3f.dll
Wideband
Spectrum
OutbandSignalPowerSpectrum.cs
NB2 Blocking
Noise Blanker
BlockNoiseBlanker.cs
NB1 Avg Noise
Blanker
AveragingNoiseBlanker.cs
LO adjust
Oscillator.cs
Main Panadapter
SignalPowerSpectrum.cs
Receiver BW
Filter
Filter.cs
S Meter
SignamMeter.cs
Wideband Spectrum EP4 data
Auto Notch Filter
InterferenceFilter
.cs
Noise Filter
NoiseFilter.cs
Demodulator
PLL.cs
Squelch
Squelch.cs
AGC
wcpagc.cs
SharpDSP DoDSPProcess

EP6 USB frame structure

EP6 data frame for a single receiver (data from radio to PC)

EP6	Sync			Command & control 5 bytes					I stream Rx1 24 bits			Q stream Rx1 24 bits			Mic audio		I stream Rx1 24 bits			Q stream Rx1 24 bits		
	0x7F	0x7F	0x7F	C0	C1	C2	C3	C4	I2	I1	I0	Q2	Q1	Q0	M1	M0	I2	I1	I0	Q2	Q1	Q0

- 512 byte frame with 504 byte payload
- 63 Microphone, I and Q samples per USB frame

EP6 USB frame structure

Example of EP6 data frame for three receivers (data from radio to PC)

EP6	Sync			Command & control 5 bytes					I stream Rx1 24 bits			Q stream Rx1 24 bits			I stream Rx2 24 bits			Q stream Rx2 24 bits		
	0x7F	0x7F	0x7F	C0	C1	C2	C3	C4	I2	I1	I0	Q2	Q1	Q0	I2	I1	I0	I0	Q2	Q1

- 512 byte frame with 500 byte payload
- 25 Microphone, I and Q samples per USB frame

EP6 USB frame structure

Continued EP6 data frame for a single receiver (data from radio to PC)

Byte 22	Byte 23	Byte 24	Byte 25	Byte 26	Byte 27	Byte 28	Byte 29	Byte 30	Byte 31		Byte 504	Byte 505	Byte 506	Byte 507	Byte 508	Byte 509	Byte 510	Byte 511
Mic audio		I stream Rx1 24 bits			Q stream Rx1 24 bits			Mic audio		etc.	I stream Rx1 24 bits			Q stream Rx1 24 bits			Mic audio	
M1	M0	I2	I1	I0	Q2	Q1	Q0	M1	M0	etc.	I2	I1	I0	Q2	Q1	Q0	M1	M0

EP6 USB frame structure

Continued example of EP6 data frame for three receivers (data from radio to PC)

Byte 21	Byte 22	Byte 23	Byte 24	Byte 25	Byte 26	Byte 27	Byte 28		Byte 500	Byte 501	Byte 502	Byte 503	Byte 504	Byte 505	Byte 506	Byte 507	Byte 508	Byte 509	Byte 510	Byte 511
I stream Rx3 24 bits			Q stream Rx3 24 bits			Mic audio		...	I stream Rx3 24 bits			Q stream Rx3 24 bits			Mic audio		Padding			
I2	I1	I0	Q2	Q1	Q0	M1	M0		I2	I1	I0	Q2	Q1	Q0	M1	M0	0	0	0	0

EP2 USB frame structure

EP2 data sent to radio from PC

	Byte 0	Byte 1	Byte 2	Byte 3	Byte 4	Byte 5	Byte 6	Byte 7	Byte 8	Byte 9	Byte 10	Byte 11	Byte 12	Byte 13	Byte 14	Byte 15
	Sync			Command & control 5 bytes					Left audio 48kHz 16 bits		Right audio 48kHz 16 bits		I stream Tx 16 bits		Q stream Tx 16 bits	
EP2	7F	7F	7F	C0	C1	C2	C3	C4	L1	L0	R1	R0	I1	I0	Q1	Q0

- 16 bit 48 kHz left and right audio for speaker jack
- 16 bit 48 kHz I signal and Q signal to be transmitted
- 512 byte frame with 504 byte payload

EP4 Wideband Spectrum frame structure

Wideband spectrum EP4 data

	Byte 0	Byte 1	Byte 2	Byte 3	Byte 4	Byte 5	Byte 6	Byte 7	Byte 8	Byte 9	Byte 10	Byte 11	Byte 12	Byte 13	Byte 14	Byte 15
	ADC 16 bit sample		ADC 16 bit sample		ADC 16 bit sample		ADC 16 bit sample		ADC 16 bit sample		ADC 16 bit sample		ADC 16 bit sample		ADC 16 bit sample	
EP4	D1	D0	D1	D0	D1	D0	D1	D0	D1	D0	D1	D0	D1	D0	D1	D0

- no sync or command & control just data 1024 bytes per Ethernet frame
- 1024 byte frame with 1024 byte payload (all data)

EP2 USB frame structure

Continued EP2 data sent to radio from PC

Byte 16	Byte 17	Byte 18	Byte 19	Byte 20	Byte 21	Byte 22	Byte 23		Byte 504	Byte 505	Byte 506	Byte 507	Byte 508	Byte 509	Byte 510	Byte 511
Left audio 48 kHz 16 bits		Right audio 48 kHz 16 bits		I stream Tx 16 bits		Q stream Tx 16 bits		...	Left audio 48 kHz 16 bits		Right audio 48 kHz 16 bits		I stream Tx 16 bits		Q stream Tx 16 bits	
L1	L0	R1	R0	I1	I0	Q1	Q0	...	L1	L0	R1	R0	I1	I0	Q1	Q0

EP4 Wideband Spectrum frame structure

Continued wideband spectrum EP4 data

Byte 16	Byte 17	Byte 18	Byte 19	Byte 20	Byte 21	Byte 22	Byte 23	Byte 24	Byte 25		Byte 1016	Byte 1017	Byte 1018	Byte 1019	Byte 1020	Byte 1021	Byte 1022	Byte 1023
ADC 16 bit sample		ADC 16 bit sample		ADC 16 bit sample		ADC 16 bit sample		ADC 16 bit sample		...	ADC 16 bit sample		ADC 16 bit sample		ADC 16 bit sample		ADC 16 bit sample	
D1	D0	D1	D0	D1	D0	D1	D0	D1	D0	...	D1	D0	D1	D0	D1	D0	D1	D0

The wideband spectrum process

The Wideband Spectrum Process starts in the DataLoop method of the EthernetDevice block. The EP4 data is bundled into the *EP4buf* buffer as either an 8 kB or a 32 kB block, (4096 or 16384 16 bit samples). Then inside the ProcessWideBandData method, the block of data is converted to + and – values and copied into identical I and Q streams because the FFT needs I and Q data centred at zero.

Next the FullBandwidthSpectrum.Process in the OutbandPowerSpectrumSignal block is called. This in turn calls the FFT routine to create a spectrum signal from the time domain input signal. As explained in the previous section, the FFT output is in two halves and each half of the buffer has to be reversed for correct display of the spectrum. The real and imaginary samples are combined to give a signal magnitude and the levels are made logarithmic to match the dB scale on the spectrum display. A separate buffer contains an averaged version of the spectrum data to provide the option to display an averaged panadapter. When completed the corrected wideband spectrum data is in the *ps_results* buffer and the averaged spectrum is in the *ps_average* buffer. Both contain either 4096 or 16,384 frequency bins containing the log scaled magnitude of the spectrum signal at different frequencies.

Next the program raises a *PSpectrumEvent* event to signal to the Kiss Konsole application that there is a chunk of wideband data to display. On seeing the event, Kiss Konsole loads the *Full Bandwidth Power Spectrum Data* buffer with the FFT spectrum data.

In another thread, Form1 responds to the timer1 'tick' and draws the data on the wideband screen display using the ComputeDisplay() method. Then it goes on with other tasks until the next 'tick' before drawing the line again for the next spectrum update.

The sample rate for the wideband spectrum is 122.88 MHz so at this stage there are 4096 or 16,384 frequency samples extending from 0 Hz to 61.44 MHz (Fs/2) in the 2nd half of the buffer. The 1st half of the buffer contains a mirror image or 'negative frequencies' which are of no use for the wideband spectrum display. We have either 4096 or 16,384 samples but the spectrum display is only 1024 pixels across and 256 pixels high. This means there are four or possibly sixteen FFT samples for each dot across the screen. The program works its way through the samples in the 2nd half of the buffer and takes the maximum value of every four samples, or the maximum value of every of every sixteen samples for a 32k buffer. The numbers could have been averaged but using the maximum is much easier and provides a sort of peak hold effect. Each pixel is scaled to a level which fits on the available display height and width. The result is 1024 numbers which are stored in a *Draw* array and eventually displayed as the wideband spectrum pixels.

The transmit process

The Transmit Process actually starts with the receive process in the HPSDRDevice block. Kiss Konsole has the ability to send audio from the microphone or 'dashes and dots' from a Morse key, plugged into the radio (Hermes board). So, the PTT signal and microphone audio, or the Morse code information, has to be obtained from the data sent from the radio to the PC in the EP6 data frames. It is processed and then sent back to the radio in the EP2 frames for transmission.

There is no facility in Kiss Konsole for VAC (virtual audio cable) operation so you can't send audio from another PC application such as digi-mode software. Remember that Kiss Konsole is really a learning tool rather than a fully capable SDR transceiver application. I am rather surprised how simple the transmit side of the application is. The modulation code is very compact.

The PTT signal and the Morse key dash and dot are sent from the radio to the PC as part of the command and control bits, (C0 bits 0-3), in every EP6 frame. All other command and control signals are only sent once in every five EP6 frames.

The audio from the microphone is sampled with an ADC in the radio and sent to the PC as a 16 bit data stream inside the EP6 data. The HPSDRDevice *Process_Data* process pulls out the mic data and eventually stores it in the TransmitFilterBuffer. But first a lot of audio processing is applied.

The mic audio is always sampled at 48 kHz, so when the receiver sample rate is also set to 48 kHz there is one stream of mic audio along with the receiver I and Q data. But when the receiver sample rate is increased to 96, 192 or 384 kHz, we still only need mic data at 48 kHz so it is simply duplicated into the higher data rate streams. At the 96 kHz sample rate the mic audio is sent twice and at 384 kHz it is sent eight times. This means in order to just get the wanted 48 kHz mic audio data, the samples must be decimated according to how fast the receiver sample rate is.

Once we have extracted the mic audio data, functions like; mic gain, VOX, bass cut and mic AGC are applied in the *ProcessorFilter.Process()* in Filter.cs, which calls FFT and creates a complex (real and imaginary) signal in the frequency domain. It completes the audio processing then a calls a reverse FFT function to return the signal back to time domain I and Q samples.

The I and Q data then goes to the *TransmitFilter.Process()* in Filter.cs, which applies the audio bandwidth filtering and a phase difference between real and imaginary signals. Just like the processor filter it does an FFT, then the filtering, then reverse FFT to return the signal back to time domain I and Q samples. It sounds complicated but what it means is we end up with audio which has had speech processing applied and has been filtered to the wanted transmit bandwidth, usually less than 3 kHz for SSB. Finally the audio is put into the *TransmitAudioRing* buffer ready to be sent to the *Data_send_core* method

Data_send_core makes the EP2 USB data frame. It inserts the sync and command & control bytes, the left and right receiver audio channels for the speakers, and the 16 bit transmit I and Q signal. But first the modulation has to be applied. The *TransmitAudioRing* buffer is read and maths is done to create FM, SSB or AM modulation.

Single sideband modulation

For upper sideband SSB, the I and Q output samples are only adjusted in level according to the microphone gain control setting 'MicScaleOut'. SSB modulation is just a matter of shifting the audio signal up to an RF frequency, so no audio processing is required. The complex signals are converted to two streams of 16 bit integers. The real signals become the I stream and the imaginary signals become the Q stream. Lower sideband SSB is the same except the real signal becomes the Q stream and the imaginary signal becomes the I stream. This inversion changes the phase between the streams from +90 to -90 degrees so that the lower sideband signal is generated the on the low side of the local oscillator frequency of the CORDIC in the FPGA.

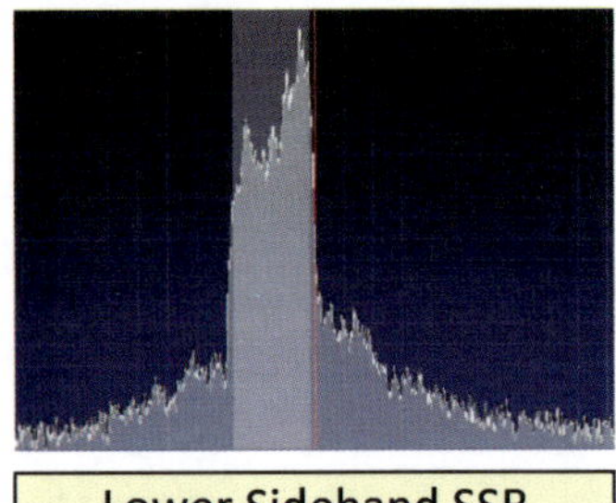

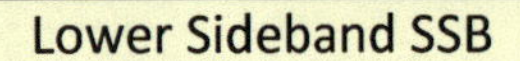

Lower Sideband SSB

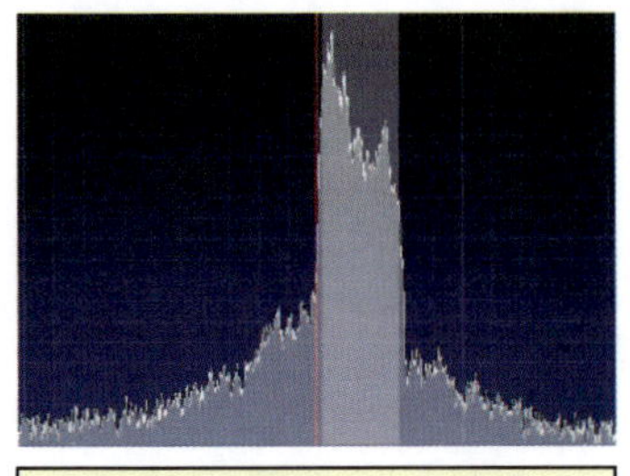

Upper Sideband SSB

Upper sideband is created using these two lines of software code.

```
I_data = (short)(MicScaleOut * SampleReal);
Q_data = (short)(MicScaleOut * SampleImag);
```

Lower sideband is created using these two lines of software code.

```
Q_data = (short)(MicScaleOut * SampleReal);
I_data = (short)(MicScaleOut * SampleImag);
```

CW modulation

CWL and CWH is CW generated by sending a tone with either lower sideband modulation or upper sideband modulation respectively. i.e. the tone will be offset either below or above the nominal transmit frequency. This is the same as all modern transceivers. As with the upper and lower sideband modulation above, changing from CWL to CWH is just a matter of swapping the signals being sent to the I and Q streams. The software looks to see if the dash or dot signal is present in the EP6 frame. If the dot or dash bit is a 1 the program sends the tone to the transmitter. When the Morse key is first pressed a raised cosine shaping factor is added to the start of the CW tone pulse and after the PTT or key is released a raised cosine shaping factor is added to the end of the CW tone pulse. This is essential to avoid 'key clicks' and excessive transmit bandwidth on CW. There is no timing involved in the CW generation in the PC. The dash and dot timing is managed by the keyer in the radio, so the dash and dot signals received in the EP6 C&C bits already have CW timing applied. If the sequencing option on Form1 is selected, the tone output is delayed for a few milliseconds after PTT is sent to allow the linear amp to stabilise.

AM AND SAM MODULATION

For AM or SAM modulation the real samples are used to generate the I stream. The imaginary signals are not used and the Q stream is set to all zeros. A value of 1 is added to the microphone sample level to shift the number range of the microphone signal from values between -1 and +1 to values between 0 and 2 The microphone level 'MicScaleOut' is reduced to 40% of whatever the mic level control is set to, so that the maximum output level is 0.4 x 2 = 0.8 which represents an 80% modulation depth. There is no other audio processing for AM or SAM modulation.

```
I_data = (short)(MicScaleOut * 0.4f * (1 + SampleReal));
Q_data = 0;
```

Setting the Q stream to 0 causes the CORDIC mixer in the FPGA to create mirrored sidebands above and below the local oscillator frequency because there is no Q stream for image cancellation. This is how the AM signal becomes double sideband.

FM MODULATION

FM modulation is much more complex! Let's just say more mathematical wizardry takes place. You can skip ahead if your head hurts already. The FM modulation is completed with only eight lines of C# code including the software loop. The actual modulation part is achieved with the four lines below.

In Kiss Konsole, FM is achieved by phase modulation. Only the real samples out of the buffer are used. The imaginary samples are not required for FM modulation. At the 48 ksps sample rate, the software calculates how much the modulating signal would change the phase of a carrier oscillator, if there was one. The code shown below creates a value that sets the phase shift that would need to be applied to each sample in order to achieve peak deviation. Kiss Konsole calls it 'cvtmod2freq'.

```
cvtmod2freq = (float)MainForm.FM_deviation * 2.0 * Math.PI / 48000.0;
```

The phase angle that a sine wave signal has to go through to complete a full cycle is $2\pi r$ Radians (360 degrees). In Kiss Konsole the maximum allowed FM_deviation is set in Form1 to 2400 Hz. The deviation is the r in the formula, it is multiplied by 2 x Pi, making $2\pi r$. The result is divided by the number of samples per second to give the maximum change of phase angle which is allowed per sample.

Using the formula and the deviation set in Form1, cvtmod2freq becomes 0.314159 Radians of phase change per sample, i.e. a maximum of exactly 18 degrees of phase change per sample. If you talk quietly the signal deviates less than 0.314159 Radians per sample. The 'oscphase' variable stores the phase modulated signal. It is like an oscillator running at 0 Hz being phase modulated by the mic audio.

```
oscphase += SampleReal * cvtmod2freq;
```

As each sample of microphone audio is read from the buffer, its level is used to modify the phase of the nominal oscillator. The phase of the oscillator is equal to its phase during the previous sample plus the level of the microphone audio (SampleReal) times the cvtmod2freq phase shift constant.

This single line of code applies the FM phase modulation to the oscillator.The larger the amplitude of the microphone audio the more phase shift is applied. But since the microphone audio is always between plus 1 and minus 1, the phase change can never exceed plus or minus 0.314159 Radians per sample. So the peak deviation is limited to +-2400 Hz. Higher audio frequencies cause a higher rate of phase change.

The I and Q signals are created using;

```
I_data = (short)(MicScaleOut * Math.Cos(oscphase));
Q_data = (short)(MicScaleOut * Math.Sin(oscphase));
```

Using the Sine and Cosine of the phase angles creates a 90 degree phase difference between the I and Q signals. MicScaleOut is the microphone level control, it just scales the samples to the correct levels.

After the modulation

After the modulation or audio processing has been applied, the I and Q samples are added to the EP2 USB frame as 16 bit words. At this stage the USB frame is complete. It is loaded into an array which is read by the Ethernet process. In the Ethernet.cs thread the *DataLoop* method waits until it has two USB frames, then it adds the sync, status, EP and sequence number bytes and sends the EP2 Ethernet packet to the radio.

The receive process

The Receive Process is the most complicated because there is a lot to do. It starts in HPSDRDevice with the *Process_Data* method. The EP6 USB data packages contain the 24 bit I and Q signals for up to seven receivers (panadapters). In Kiss Konsole only one panadapter can be used, so this discussion will only cover one receiver demodulation and one panadapter spectrum display. Software which supports more than one receiver duplicates and reuses blocks of code for each panadapter.

Process_Data checks that data is being received and sets the green Sync indicator. It decodes the command and control information from each frame and sets variables to be read by the GUI interface and other code blocks, then it scales the I and Q 24 bit numbers and puts them as large 'float' numbers into the *SignalBuffer* buffer. All of the rest of the code in *Process_Data* deals with the audio signal from the microphone, so when 1024 samples have been accumulated the *DoDSPProcess* method in the Receiver.cs DSP block is called.

The following explanation of the DSP process will be much easier to understand if you look at the functional block diagram.

SharpDSP digital signal processing

DoDSPProcess in Receiver.cs runs the receiver DSP process in SharpDSP. The SharpDSP code blocks are the green boxes on the functional block diagram. The following methods or 'calls' are performed in turn.

Blocking noise blanker (NB2)

If the blocking noise blanker (NB2) is selected, *DoDSPProcess* calls BlockNoiseBlanker.cs. Depending on the threshold setting, the blocking noise blanker can completely remove noise spikes which are bigger than the average signal level, but it has no effect on noise spikes that are below the average signal level.

The noise blanker does not act as a clipper it mutes the signal completely where there is a large noise spike. Using NB2 introduces a, 2 byte delay (latency) to the received signal. When it acts, it blanks the signal to zero for 7 bytes. Hopefully by then the noise impulse has passed. At higher sample rates the data stream is faster so the latency and mute period is shorter.

Averaging noise blanker (NB1)

If the averaging noise blanker (NB1) is selected, *DoDSPProcess* calls AveragingNoiseBlanker.cs. The averaging noise filter also works on signals greater than the average signal level and the threshold level set in the GUI software. It works by replacing the noise spike with a level equal to an average of recent signal samples. The theory is that the period of the noise spikes will be much shorter than the wanted signal, so it is quite effective on impulse noise. It does not completely remove the noise spike and if the threshold is set too aggressively it can cause unwanted distortion to the wanted signal.

Oscillator block

Oscillator.cs - I am 99% sure that this block is unused in Kiss Konsole. The LO Frequency is set to 0 by Form1 and is not changed. I think it is used to provide a variable offset for some types of receivers.

The code block generates a tone frequency, calculates its phase and magnitude at each sample time and then offsets the signal phase and amplitude of each signal sample by that amount. In some receiver designs the local oscillator in the receiver front end can only be adjusted in large steps like 10 kHz. The Oscillator.cs code allows you to move the spectrum and receiver around in smaller steps. The CORDIC in Hermes can be adjusted in very small increments so the Oscillator.cs code is not required.

Panadapter display

The PowerSpectrumSignal *process()* in SignalPowerSpectrum.cs creates the panadapter display. First the Windowing function is applied to a copy of the received data. The Windowing FIR filter is predefined using an FFT process. Then the FFT forward process is called to create a spectrum signal from the time domain input signal. As explained in the FFT section, the FFT output is in two halves and each half of the buffer has to be reversed for correct display of the spectrum. The real and imaginary samples are combined to give a signal magnitude and the levels are made logarithmic to match the dB scale on the spectrum display. A separate buffer contains an averaged version of the spectrum data to provide the option to display an averaged panadapter.

When completed. The corrected spectrum data is in the *ps_results* buffer and the averaged spectrum is in the *ps_average* buffer. Both contain 2048 FFT bins containing the log scaled magnitude of the spectrum signal at different frequencies. The program raises a *PSpectrumEvent* event to signal to the Kiss Konsole application that there is a chunk of panadapter data to display. On seeing the event, Kiss Konsole loads the *PowerSpectrumData* buffer with the FFT spectrum data.

In another thread, Form1 responds to the timer1 'tick' and draws the data on the panadapter screen display using the ComputeDisplay() method. Then it goes on with other tasks until the next 'tick' before drawing the line again for the next spectrum update.

There are 2048 samples in the FFT data but the spectrum display is only 1024 pixels across, so every second sample is used for the display.

The level for each pixel is scaled so that the spectrum display will fit on the available display height and width and the spectrum line is drawn. The spectrum is refreshed at between 1 and 50 lines per second, as per the FPS (frames per second) setting on the Setup form.

The waterfall display

Generating the waterfall display is relatively easy and no additional data from the radio is required. The spectrum line we just drew contains everything we need for the waterfall display. Inside the loop which generates the spectrum pixels, an RGB (red green blue) colour is calculated for each of the 1024 spectrum pixels. The higher the level of the signal the brighter and more active the colour of the pixel. When the loop terminates and the panadapter spectrum line has been drawn, the *rawData* array holds RGB and brightness information for 1024 waterfall pixels across the screen. The existing waterfall is a 256x1024 pixel bitmap image, so the line to be added is defined as a rectangle and then added to the existing bit map to form a new line on the image. This means the waterfall does not fill by scanning across the display like the panadapter spectrum, it fills a complete line at a time. The number of frames per second is adjustable on the setup form. If you set it to 1 you can see each spectrum and waterfall line update once per second and it will take 256 seconds for the waterfall screen to scroll all the way down.

The maximum scroll speed for the waterfall that can be obtained from the 48 ksps data stream is about 5.4 seconds but my system tops out at around 8 seconds to fill the 256 line display. This equivalent to setting 32 frames per second. On my PC, any faster setting gives no increase in display speed.

Receiver filter for the audio pass band

So far we have done noise blanking if selected and displayed the spectrum and waterfall images if they are selected. Filter.cs is the start of creating the audio we listen to. We want to apply a filter so we only end up with the spectrum inside the wanted audio bandwidth, for example the 2.4 to 3 kHz wide part of the spectrum we want for SSB. Or the 100 to 500 Hz bandwidth for receiving a CW signal. Applying these filters is done on frequency domain signals, so the first step is to do an FFT transformation by calling the FFT in the DSPBuffer.cs module. The filter is applied using mathematical magic in the *DoConvolution()* method and then the signal is turned back into time domain samples using reverse FFT. Convolution in the frequency domain is equivalent to applying a mixer. The wanted bandwidth is selected by multiplying the frequency domain spectrum against the pre-determined filter shape. When converted back to the time domain the signals have levels between -1 and 1.

By the way, the filter bandwidth is drawn onto the background image of the spectrum display as a part of the Form1 code and not as a part of this routine. It is updated as a result of changes to the audio bandwidth settings, such as clicking the 2.6 kHz button or dragging the filter with the mouse.

Receiver audio pass-band filter (Filter.cs)

These pictures show the action of the Filter on the frequency domain signals. This is a real upper sideband SSB signal captured from Kiss Konsole and charted in Excel. You can see the input signals, the filter shape and the output signals. The spectrum view right at the bottom was taken using the magnitude of the output signals $\sqrt{I^2+Q^2}$.

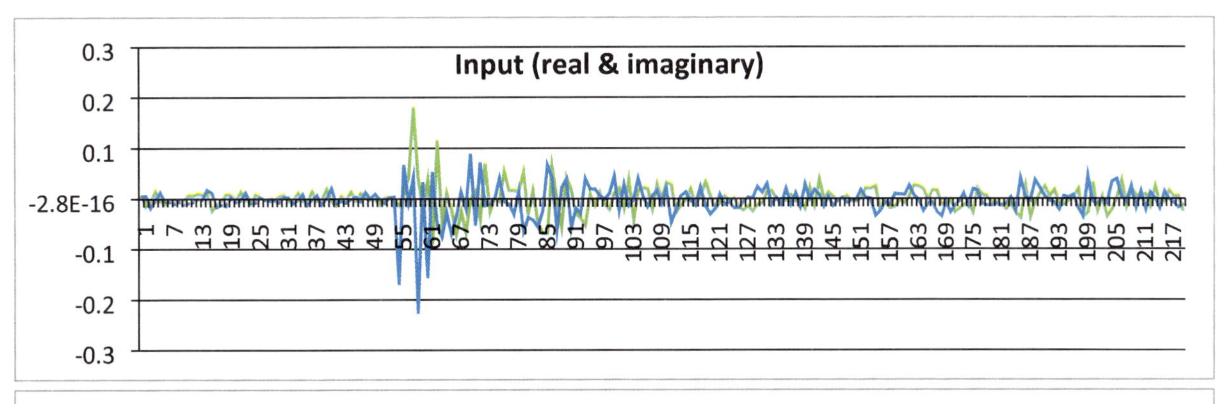

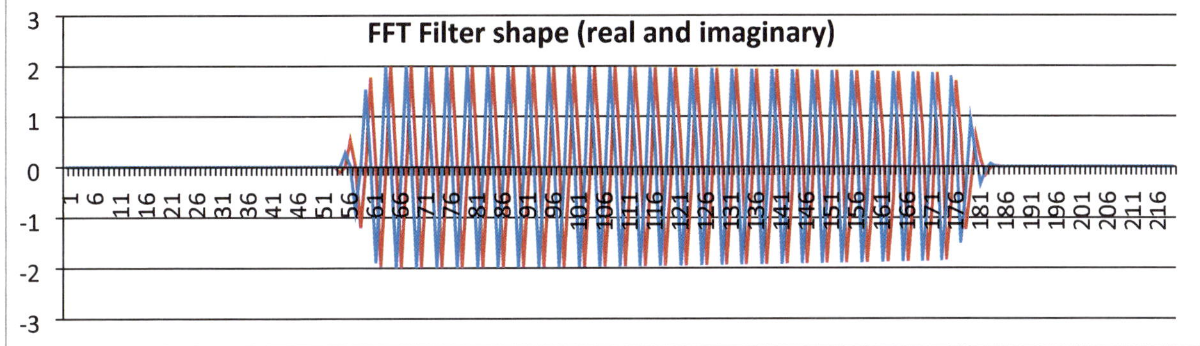

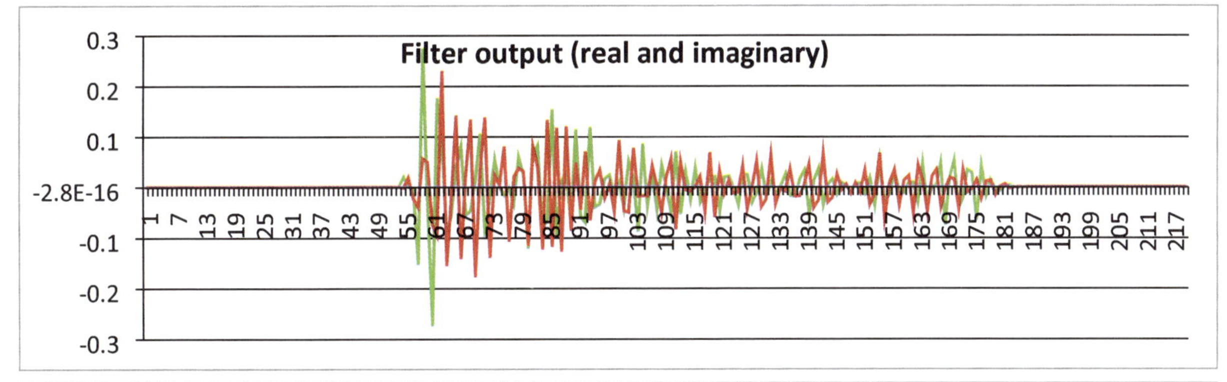

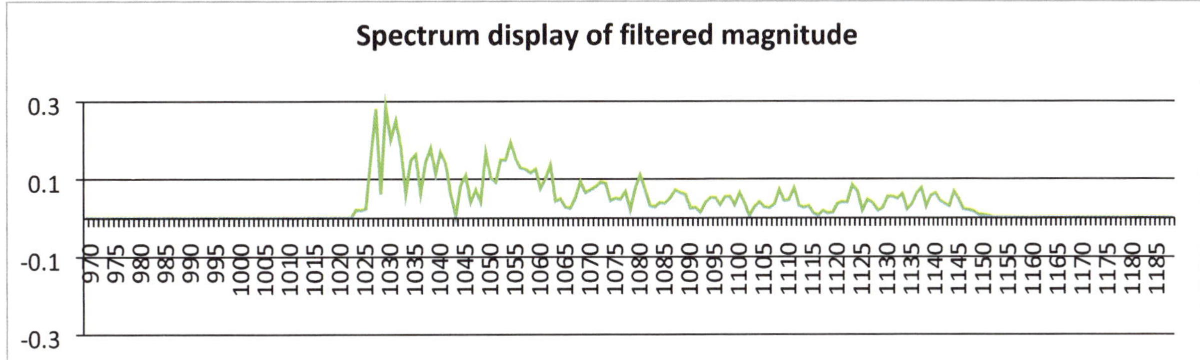

The S meter

SignalMeter.cs is called to do the S meter calculation but not its display. It goes to the DSP process and adds together the power of the samples inside the filtered spectrum. Then it converts the result to a logarithmic value so that the signal level can be displayed in dBm and updates an 'instantaneous received signal level' variable and an 'average signal level' variable. The averaged variable is a moving average of the last 10 samples. A strobe counter is used in the same way as the one in the spectrum display code so the meter update rate is the same at all sample rates (i.e between 48 ksps and 384 ksps). In my copy of Kiss Konsole, I changed the average to 1000 samples in order to slow down the text display of the signal level. In my opinion the average of 10 samples is much too fast to read comfortably, but I guess it is a matter of personal preference. When the S meter calculation is complete an event is raised to let the Form1 thread, know that the S meter display and text can be updated.

AGC

Next AGC is applied to the audio levels using Wcpagc.cs. There are three different AGC modules in SharpDSP, but the one used in Kiss Konsole is wcpagc by Warren Pratt, NR0V. AGC is one of the largest code blocks in the whole application. There is a choice of AGC off, long, slow, medium, or fast, with adjustable AGC slope, decay and hang time.

The AGC code is very maths intensive because there are so many variable features. I am not going to even attempt to analyse it. AGC works on the signal levels in the time domain so no FFT is required.

Squelch

The code in Squelch.cs is not used by Kiss Konsole. It is different to the squelch function in Form1 which has the option of applying squelch based on the signals inside the receiver pass-band (around 2.5 kHz for SSB), or an alternate squelch method based on the level across the whole panadapter. The code in Form1 calculates the maximum signal across the whole panadapter using the numbers that are used to create the panadapter display. For squelch based on the signal inside the receiver pass-band, it uses the maximum of the signals fed to the S meter. If the maximum level is less than the squelch control setting the receiver audio will be squelched. To squelch the audio it simply turns the volume control down to zero.

Demodulation

Next PLL.cs is called to do the demodulation of AM, SAM (synchronous AM) and FM signals.

SSB signals don't need any demodulation processing because the samples have already been converted down to an audio bandwidth by the filter method.

The **AM** demodulator is an envelope detector much like the diode and capacitor in a crystal set. The magnitude of the signal is calculated using code in the DSPBuffer block from the I and Q signals using the vector sum ($\sqrt{I^2 + Q^2}$). A moving average of 1000 samples is used to find the DC value and the signal is adjusted in level so that it is centred at 0 Volts. The input signal is all positive values. The output signal is smoothed by averaging 10 samples and copied to the CPX buffer. The actual AM demodulation is completed using a single line of software. It calculates the magnitude of the IQ sample using $\sqrt{I^2+Q^2}$.

```
AM = (float)Math.Sqrt(cpx[index].real * cpx[index].real +cpx[index].imaginary * cpx[index].imaginary);
```

The **SAM** and **FM** demodulation methods are very similar to each other. Both use a phase locked loop to determine the centre point of the signal. The demodulated signal is the amount the signal varies against the centre point. Because SAM is a type of AM modulation, the SAM demodulator detects the magnitude as well as the phase change of the signal this means that the combined level of both sidebands is being detected. The FM demodulator detects the phase change of the input signal and not the magnitude. These methods are relatively complicated taking around 20 lines of software code.

Noise reduction filter

If the NR (noise reduction) filter is selected, a call is made to NoiseFilter.cs. The noise filter is an adaptive interference canceller. It finds a periodic (repeating) signal like speech and cancels out incoherent signals such as noise. The data stream is passed through a delay line and an adjustable FIR (finite impulse response) filter, which is a band pass filter that automatically moves to the signal we want to keep, or in the case of a notch filter eliminate. The delayed signal becomes a kind of reference signal at the periodic frequency. It is subtracted from the current samples creating an error level. After some time the filter coefficient has adjusted to progressively reduce the error signal, meaning that the delayed signal is as close as possible, in phase with the current samples. When the error signal is at its smallest, the wanted periodic speech is enhanced and the noise is suppressed. The technique uses a least mean square algorithm developed by Widrow and Hoff in the late 1950s. These adaptive filters work in the time domain, no FFT is required.

Notch filter

If the ANF (auto notch filter) is selected, a call is made to InterferenceFilter.cs. The notch filter is very similar to the noise filter. In fact the software code is almost identical although the reference variables have different values. But in this case the unwanted signal is periodic constant signals like unwanted carriers and 'birdies' and the wanted signal is the less periodic speech. So this time we output the more incoherent 'error' signal rather than the coherent interference signal. The auto notch filter can't be used when you want to receive relatively constant signals like PSK or RTTY or even CW because it would attempt to cancel the wanted signal.

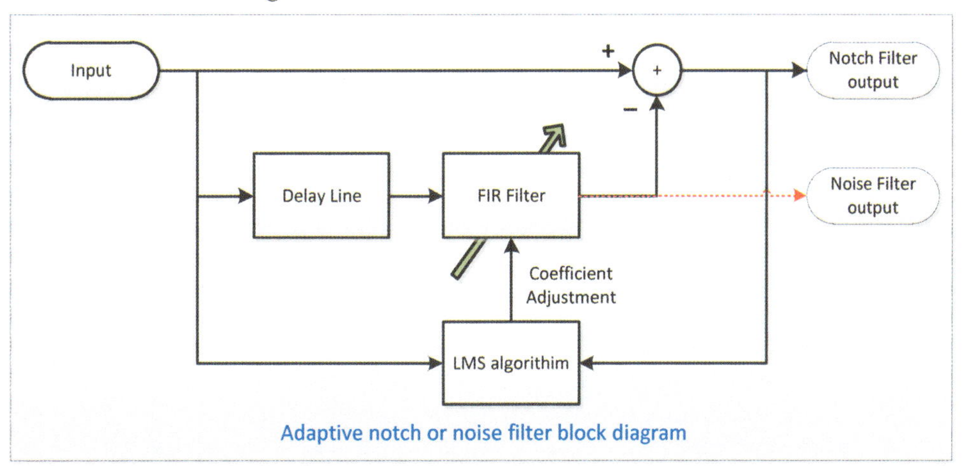

Adaptive notch or noise filter block diagram

Dr Widrow discovered a way to alter the FIR filter coefficients "on the fly" by using the adaptive LMS (least mean squares) method.

Changing the filter coefficients adjusts the band pass filter so that the periodic interference signal is notched out. Or for the noise filter, the output signal is inside the band pass filter and the noise outside the band pass filter is filtered out.

AUDIO OUTPUT

Output.cs is used to decide whether the audio should be sent to the left speaker, the right speaker, both speakers as stereo, or mixed together for both speakers. In Kiss Konsole there is no balance control so the left and right volume levels are both set with the volume slider control on Form1. A module in DSPBuffer.cs scales the signal level to the wanted volume level. If the output is set for both speakers, both the real (right) and imaginary (left) streams are set to the same level. If the right speaker is selected the left stream is set to all zeros. If the left speaker is selected the right stream is set to all zeros. The mixed output will have no effect on Kiss Konsole because the audio is the same on both channels. In another application it could be used when you have the output of RX1 on one audio channel and the output of RX2 on the other. A slide control would allow the operator to put some of the left audio onto the right channel and some of the right audio onto the left channel. The binaural mode is the same as the 'both' mode except the signals are not made the same on both channels. So if you wanted to add a balance control or to listen to different receivers with one on the left and the other on the right, the binaural or mixed mode would be used. In Kiss Konsole selecting binaural has no effect.

THE END OF THE RECEIVER PROCESS

As a part of the *DoDSPProcess* the panadapter and waterfall data has been produced by SignalPower Spectrum.cs and the data for the wideband spectrum display has been produced by the OutbandPowerSpectrum block. The display of these is handled by code in the Form1 software block.

After the *DoDSPProcess* has been completed, control returns to Receiver.cs and then to the HPSDRDevice module.

The receiver audio is bundled into the *AudioRing* buffer to be sent to the speakers on the radio via the EP2 Ethernet packets. The audio sent to the radio is at 48 ksps, but the audio coming out of the DSP process is at the same rate as the receiver sample rate. So the audio may have to be decimated before it is inserted into the *AudioRing* buffer and eventually added into the EP2 frame.

- If the EP6 frames from the radio are at the 48 ksps sample rate then one audio frame is sent to the radio for each EP6 frame received.
- If the EP6 frames from the radio are at the 96 ksps sample rate then one audio frame is sent to the radio for every two EP6 frames received.
- If the EP6 frames from the radio are at the 192 ksps sample rate then one audio frame is sent to the radio for every four EP6 frames received.
- If the EP6 frames from the radio are at the 384 ksps sample rate then one audio frame is sent to the radio for every eight EP6 frames received.

The Command and Control system

The Command and Control information is sent to the radio in the five 'CC' bytes at the start of each of the EP2 frames and received from the radio in the in the five 'CC' bytes at the start of each of the EP6 frames.

The frame structure is shown on the Command and Control bytes drawing. Some of the data sent and received is not used by the Hermes board and Kiss Konsole as it is intended for earlier OpenHPSDR equipment. The CC information is extracted in the *Process_Data* method and inserted in the *Data_send_core* method of the HPSDRDevice.cs module, using and setting variables set on the Form1 and Setup forms.

Command and Control bytes received from Hermes in the EP6 stream

The PTT signal and the Morse key dash and dot are sent from the radio to the PC in every EP6 frame. All of the other EP6 command and control signals are only sent once in every five frames.

The CC data received from the radio repeats every 5 frames. The C0 byte acts as the address so the software knows whether it is seeing command and control information from frame 0, 1, 2, 3 or 4.

C0 byte	C1 byte	C2 byte	C3 byte	C4 byte
PTT & CW bits plus CC frame number = 0	Bits to indicate ADC overload, user IO lines, frequency changed	Unused by Hermes	Unused by Hermes	Hermes Software serial number
PTT & CW bits plus CC frame number = 1	Hermes forward power MSB (AIN5)	Hermes forward power LSB (AIN5)	Alex forward power MSB (AIN1)	Alex forward power LSB (AIN1)
PTT & CW bits plus CC frame number = 2	Alex reverse power MSB (AIN2)	Alex reverse power MSB (AIN2)	External metering AIN3 MSB	External metering AIN3 MSB
PTT & CW bits plus CC frame number = 3	External metering AIN4 MSB	External metering AIN4 MSB	+12 Volt rail meter MSB	+12 Volt rail meter LSB
PTT & CW bits plus CC frame number = 4	Unused by Hermes	Unused by Hermes	Unused by Hermes	Unused by Hermes

Case 0 (C1 to C4 bytes when the C0 byte = 0)

In Kiss Konsole only three of the four user input lines are read by the code. Also the ADC overload bit is being read every frame when it is only valid every 5th frame. This does not cause a problem because the C1 byte when C0 = 1, 2 or 3 is only used for transmit information. The serial numbers are only read once when Kiss Konsole starts since there is no point reading the same thing over and over.

The wideband spectrum is set initially to the larger 32kB buffer size and there is a version check to ensure that Hermes is running at least V1.8 of the FPGA firmware. These settings are also only read once, when Kiss Konsole starts.

Case 1 (C1 to C4 bytes when the C0 byte = 1)

Kiss Konsole gets the forward power for both Hermes and Alex, i.e. the 500 mW Hermes output and the 100W power amp output. But only the Alex (100W) output is displayed by Kiss Konsole. The 12 bit reading is scaled from 0-4095 (000-FFF) to 0-3.3 Volts and converted from a voltage to a power reading in Watts (0-121W).

Case 2 (C1 to C4 bytes when the C0 byte = 2)

Kiss Konsole gets the reverse power reading at the Alex filter, i.e. at the power amp output. The reading is scaled from 0-4095 to 0-3.3 Volts and converted from a voltage to a power reading in Watts (0-121W).

The routine also gets the AIN3 voltage. It is a 0-3 V external metering point on pin 12 of the Hermes J16 I/O connector which is not used by Kiss Konsole.

Case 3 (C1 to C4 bytes when the C0 byte = 3)

Kiss Konsole gets the 12 V DC supply rail voltage from the Hermes board. The reading is scaled to show between 0 Volts and 22 Volts. It is available as a variable on Form1 but it is not used except as a diagnostic test.

The routine also gets the AIN4 voltage. It is a 0-3 V external metering point on pin 11 of the Hermes J16 I/O connector which is not used by Kiss Konsole.

Case 4 (C1 to C4 bytes when the C0 byte = 4)

There is no case 4 in the Kiss Konsole software because the data in these bytes is not relevant to Hermes.

COMMAND AND CONTROL BYTES SENT TO HERMES IN THE EP2 STREAM

The MOX (PTT from software) signal is sent to the radio in every EP2 frame. All of the other EP2 command and control signals are only sent once in every ten frames.

The CC data sent to the radio repeats every 10 frames. The C0 byte acts as the address so the radio (Hermes board) knows which command and control information frame it is getting.

C0 byte	C1 byte	C2 byte	C3 byte	C4 byte
MOX bit plus CC frame number = 0	Config, mic/line Fs Sample speed 00= 48 ksps 01= 96 ksps 10 = 192 ksps 11 = 384 ksps	Class E mode (0) and six open collector output states (1-7)	Alex ATT (0-1) Preamp (2) Dither (3) Random (4) Rx Ant (5-6) Rx Out (7)	Tx Ant (0-1) Duplex/simplex (2) 1-7 receivers (3-5) Timestamp (6) Common (7)
MOX bit plus CC frame number = 1	NCO (numerically controlled oscillator) frequency - Transmit frequency			
MOX bit plus CC frame number = 2	NCO (numerically controlled oscillator) frequency – Receiver 1 frequency			
MOX bit plus CC frame number = 3	NCO (numerically controlled oscillator) frequency – Receiver 2 frequency			
MOX bit plus CC frame number = 4	NCO (numerically controlled oscillator) frequency – Receiver 3 frequency			
MOX bit plus CC frame number = 5	NCO (numerically controlled oscillator) frequency – Receiver 4 frequency			
MOX bit plus CC frame number = 6	NCO (numerically controlled oscillator) frequency – Receiver 5 frequency			
MOX bit plus CC frame number = 7	NCO (numerically controlled oscillator) frequency – Receiver 6 frequency			
MOX bit plus CC frame number = 8	NCO (numerically controlled oscillator) frequency – Receiver 7 frequency			
MOX bit plus CC frame number = 9	Transmit drive level 00-FF (0-7)	Mic boost (0) Mic / line input (1) N/A Hermes (2) N/A Hermes (3) N/A Hermes (4) SelFilt (5) HPF active (6) VNA mode (7)	13 MHz HPF (0) 20 MHz HPF (1) 9.5 MHz HPF (2) 6.5 MHz HPF (3) 1.5 MHz HPF (4) Bypass HPF (5) 6m LNA enable (6) TR enable (7)	30/20m LPF (0) 60/40m LPF (1) 80m LPF (2) 160m LPF (3) 6m LPF (4) 10/12m LPF (5) 15/17m LPF (6) unused (7)

- VNA = vector network analyser
- TR is the transmit / receive change over relay

Case 0 (C1 to C4 bytes when the C0 byte = 0)

The AlexState is decoded for the band currently in use. It contains;

- 2 bits for the Receiver attenuator setting (00 = 0 dB, 01 = 10 dB, 10 = 20 dB, 11 = 30 dB),
- 2 bits for the Alex Transmit relay on transmit (00 = Ant1, 01 = Ant2, 10 = Ant3)
- 3 bits for the receive antenna (000 = none, 001 = Rx1, 010 = Rx2, 11 = XVTR)
- 2 bits for the Alex Transmit relay on receive (00 = Ant1, 01 = Ant2, 10 = Ant3) used in the 'normal' condition i.e. when a receive only antenna has not been selected
- The C1 byte:
  - Bits 0-1 are set to the wanted sample speed (00=48ksps, 01=96ksps, 10=192ksps, 11=384ksps)
  - Bit 2-3 are set to 10 to show Hermes ("Mercury") board is using a 10 MHz clock
  - Bit 4 is set to 1 to indicate 122.88 MHz clock speed
  - Bit 5-6 are set to 10 to indicate that a Hermes ("Mercury") board is present
  - Bit 7 is not set for Hermes
  - I believe that only the sample speed is actually used by the Hermes board.
- The C2 byte:
  - Bit 0 is used to indicate the transmitter mode (1= class E, 0 = all other modes) in Kiss Konsole it is always 0
  - Bits 1-7 are used to request settings for the open collector outputs. They are currently switched off in Form1 but can be enabled by modifying the Kiss Konsole code.
- The C3 byte:
  - Bits 0-1 set the Alex attenuator (00 = 0 dB, 01 = 10 dB, 10 = 20 dB, 11 = 30 dB)
  - Bit 2 sets the preamp (0 off, 1 on)
  - Bit 3 sets ADC dither (0 off, 1 on) – not used in Kiss Konsole
  - Bit 4 sets ADC random (0 off, 1 on) – not used in Kiss Konsole
  - Bits 5-6 sets the Alex Rx antenna to one of the three receiver only ports (01 = Rx1, 10 = Rx2, 11 = XVTR) instead of one of the three main antenna ports. The C4 byte selects which of the three main ports to use.
  - Bit 7 Alex Rx out, sets the receiver antenna either the main antennas on the Alex board or to the receive only ports (0= main antennas, 1 receive only antennas).
- The C4 byte:
  - Bits 0-1 controls the Alex Tx relay to select the antenna on transmit. On receive it is sent the receive antenna, (00 = Ant1, 01 = Ant2, 10 = Ant3).
  - Bit 2 controls duplex mode (1 = full duplex, 0 = simplex)
  - Bits 3-5 set the number of receivers. In Kiss Konsole there is only one receiver so these bits are always 000
  - Bit 6 is 'Timestamp' - unused
  - Bit 7 is 'Common' - unused

Case 1 (C1 to C4 bytes when the C0 byte = 1)

Sends the 4 byte transmit (and normal Rx1) frequency

Case 2 (C1 to C4 bytes when the C0 byte = 2)

If duplex mode is selected the software sends the duplex receiver frequency. In Kiss Konsole the duplex frequency is set to the same frequency as the normal frequency

Case 3 (C1 to C4 bytes when the C0 byte = 9)

- The C1 byte:
  - The whole byte (bits 0-7) is used to send the transmit drive level
- The C2 byte:
  - Bit 0 is used to set mic gain boost by 20 dB (1 = on, 0 = off). Mic gain boost cannot be selected if line input is selected
  - Bit 1 is used to set mic or line input
  - Bits 2-7 are unused by Kiss Konsole
- The C3 byte:
  - The C3 byte is used to force HPF filter and 6m LNA switching but this function is not yet implemented in Kiss Konsole so the high pass filter always operates on 'Auto' mode.
- The C4 byte:
  - The C3 byte is used to force LPF filter switching but this function is not yet implemented in Kiss Konsole so the Alex filter board always operates on 'Auto' mode.

Case 4 (C1 to C4 bytes when the C0 byte = 3, 4, 5, 6, 7, or 8)

There is no case 4 because Kiss Konsole only provides one receiver (panadapter) so all zeros data is sent on CC frames 3-8.

Command and Control bytes received from Hermes in the EP6 stream

	C0 Byte (8 bits)							
Bit	*7*	*6*	*5*	*4*	*3*	*2*	*1*	*0*
	Command & control					Dot	Dash	PTT
CC 0	0	0	0	0	0	x	x	x
CC 1	0	0	0	0	1	x	x	x
CC 2	0	0	0	1	0	x	x	x
CC 3	0	0	0	1	1	x	x	x
CC 4	0	0	1	0	0	x	x	x

	C1 Byte (8 bits)							
Bit	*7*	*6*	*5*	*4*	*3*	*2*	*1*	*0*
CC 0	0	Freq changed	Cyclops PLL	IO 4	IO 3	IO 2	IO 1	ADC overload
CC 1	Bits 15-8 of Hermes / Penelope forward power (AIN5) (12 bits)							
CC 2	Bits 15-8 of Alex / Apollo reverse power (AIN2) (12 bits)							
CC 3	Bits 15-8 of Hermes / Penelope (AIN4) (12 bits)							
CC 4	Mercury 1 software version 0-127							M1 ADC

	C2 Byte (8 bits)							
Bit	*7*	*6*	*5*	*4*	*3*	*2*	*1*	*0*
CC 0	Mercury Software Serial No. (0 if Hermes)							
CC 1	Bits 7-0 of Hermes / Penelope forward power (AIN5) (12 bits)							
CC 2	Bits 7-0 of Alex / Apollo reverse power (AIN2) (12 bits)							
CC 3	Bits 7-0 of Hermes / Penelope (AIN4) (12 bits)							
CC 4	Mercury 2 software version 0-127							M2 ADC

	C3 Byte (8 bits)							
Bit	*7*	*6*	*5*	*4*	*3*	*2*	*1*	*0*
CC 0	Penelope Software Serial No. (0 if Hermes)							
CC 1	Bits 15-8 of Alex / Apollo forward power (AIN1) (12 bits)							
CC 2	Bits 15-8 of Hermes / Penelope (AIN3) (12 bits)							
CC 3	Bits 15-8 of Hermes / Penelope (AIN6) (13.8V on Hermes) (12 bits)							
CC 4	Mercury 3 software version 0-127							M3 ADC

	C4 Byte (8 bits)							
Bit	*7*	*6*	*5*	*4*	*3*	*2*	*1*	*0*
CC 0	OZY / Hermes Software serial No.							
CC 1	Bits 7-0of Alex / Apollo forward power (AIN1) (12 bits)							
CC 2	Bits 7-0 of Hermes / Penelope (AIN3) (12 bits)							
CC 3	Bits 7-0 of Hermes / Penelope (AIN6) (13.8V on Hermes) (12 bits)							
CC 4	Mercury 4 software version 0-127							M4 ADC

Command and Control bytes sent to Hermes in the EP2 stream

	C0 Byte (8 bits)							
Bit	*7*	*6*	*5*	*4*	*3*	*2*	*1*	*0*
CC	Command & control							MOX
CC 0	0	0	0	0	0	0	0	x
CC 0	0	0	0	0	0	0	0	x
CC 1	0	0	0	0	0	0	1	x
CC 2	0	0	0	0	0	1	0	x
CC 3	0	0	0	0	0	1	1	x
CC 4	0	0	0	0	1	0	0	x
CC 5	0	0	0	0	1	0	1	x
CC 6	0	0	0	0	1	1	0	x
CC 7	0	0	0	0	1	1	1	x
CC 8	0	0	0	1	0	0	0	x
CC 9	0	0	0	1	0	0	1	x

	C1 Byte (8 bits)							
Bit	*7*	*6*	*5*	*4*	*3*	*2*	*1*	*0*
CC 0	Mic source	Config		122.8 MHz	10 Mhz ref		Fs 48-384 kHz	
CC 1	NCO (Numeric controlled oscillator) frequency in Hz for Transmitter (32 bits)							
CC 2	NCO (Numeric controlled oscillator) frequency in Hz for Receiver 1 (32 bits)							
CC 3	NCO (Numeric controlled oscillator) frequency in Hz for Receiver 2 (32 bits)							
CC 4	NCO (Numeric controlled oscillator) frequency in Hz for Receiver 3 (32 bits)							
CC 5	NCO (Numeric controlled oscillator) frequency in Hz for Receiver 4 (32 bits)							
CC 6	NCO (Numeric controlled oscillator) frequency in Hz for Receiver 5 (32 bits)							
CC 7	NCO (Numeric controlled oscillator) frequency in Hz for Receiver 6 (32 bits)							
CC 8	NCO (Numeric controlled oscillator) frequency in Hz for Receiver 7 (32 bits)							
CC 9	Hermes / Pennylane drive level 00 - FF							

	C2 Byte (8 bits)							
Bit	*7*	*6*	*5*	*4*	*3*	*2*	*1*	*0*
CC 0	Open collector outputs							TX Mode
CC 1	NCO (Numeric controlled oscillator) frequency in Hz for Transmitter (32 bits)							
CC 2	NCO (Numeric controlled oscillator) frequency in Hz for Receiver 1 (32 bits)							
CC 3	NCO (Numeric controlled oscillator) frequency in Hz for Receiver 2 (32 bits)							
CC 4	NCO (Numeric controlled oscillator) frequency in Hz for Receiver 3 (32 bits)							
CC 5	NCO (Numeric controlled oscillator) frequency in Hz for Receiver 4 (32 bits)							
CC 6	NCO (Numeric controlled oscillator) frequency in Hz for Receiver 5 (32 bits)							
CC 7	NCO (Numeric controlled oscillator) frequency in Hz for Receiver 6 (32 bits)							
CC 8	NCO (Numeric controlled oscillator) frequency in Hz for Receiver 7 (32 bits)							
CC 9	VNA	HPF	Select Filter	Tune	Tuner	Filter	Mic/line in	Mic boost

Command and Control bytes sent to Hermes in the EP2 stream

	C3 Byte (8 bits)							
Bit	*7*	*6*	*5*	*4*	*3*	*2*	*1*	*0*
CC 0	Rx output	Rx Ant		Random	Dither	Preamp	Alex ATT	
CC 1	NCO (Numeric controlled oscillator) frequency in Hz for Transmitter (32 bits)							
CC 2	NCO (Numeric controlled oscillator) frequency in Hz for Receiver 1 (32 bits)							
CC 3	NCO (Numeric controlled oscillator) frequency in Hz for Receiver 2 (32 bits)							
CC 4	NCO (Numeric controlled oscillator) frequency in Hz for Receiver 3 (32 bits)							
CC 5	NCO (Numeric controlled oscillator) frequency in Hz for Receiver 4 (32 bits)							
CC 6	NCO (Numeric controlled oscillator) frequency in Hz for Receiver 5 (32 bits)							
CC 7	NCO (Numeric controlled oscillator) frequency in Hz for Receiver 6 (32 bits)							
CC 8	NCO (Numeric controlled oscillator) frequency in Hz for Receiver 7 (32 bits)							
CC 9	TR enable	6m LNA	Bypass HPF	1.5 MHz HPF	6.5 MHz HPF	9.5 MHz HPF	20 MHz HPF	13 MHz HPF

	C4 Byte (8 bits)							
Bit	*7*	*6*	*5*	*4*	*3*	*2*	*1*	*0*
CC 0	common	timestamp	No of receivers 1to7			Duplex	Tx Antenna	
CC 1	NCO (Numeric controlled oscillator) frequency in Hz for Transmitter (32 bits)							
CC 2	NCO (Numeric controlled oscillator) frequency in Hz for Receiver 1 (32 bits)							
CC 3	NCO (Numeric controlled oscillator) frequency in Hz for Receiver 2 (32 bits)							
CC 4	NCO (Numeric controlled oscillator) frequency in Hz for Receiver 3 (32 bits)							
CC 5	NCO (Numeric controlled oscillator) frequency in Hz for Receiver 4 (32 bits)							
CC 6	NCO (Numeric controlled oscillator) frequency in Hz for Receiver 5 (32 bits)							
CC 7	NCO (Numeric controlled oscillator) frequency in Hz for Receiver 6 (32 bits)							
CC 8	NCO (Numeric controlled oscillator) frequency in Hz for Receiver 7 (32 bits)							
CC 9	spare	15/17m LPF	10/12m LPF	6m LPF	160m LPF	80m LPF	60/40m LPF	30/20m LPF

Glossary of abbreviations and acronyms

Term	Description
A/D	Analog to digital
ADC	Analog to digital converter or analog to digital conversion
AF	Audio frequency
BDR	Blocking dynamic range receiver test
Bin	Narrow band of frequencies. (The FFT process creates thousands of narrow bins)
bit	Binary value 0 or 1
Byte	Eight binary bits
C#	C Sharp – computer language (a development of C, C+, and C++)
CAT	Computer aided transceiver - text strings used to control a ham radio transceiver from a computer program
CC	Command and Control bytes
CFIR	Compensating finite impulse response – filter type with a rise at the high frequency end to compensate for the roll off of the successive CIC filters in the FPGA
CIC	Cascaded integrated comb filters used in the FPGA to remove alias frequencies during the decimation process
CPU	Central processing unit - usually the microprocessor in a PC or other computer
D/A	Digital to analog
DAC	Digital to analog converter or digital to analog conversion
dB	Decibel a way of representing numbers in a logarithmic scale. Decibels must always be referenced to a fixed value often a Volt (dBV), milliwatt (dBm), or carrier level (dBc). Decibels are also used to represent logarithmic units of gain or loss.
DDC	Digital down conversion - the receiver part of a direct sampling SDR
DDS	Direct digital sampling (SDR) or when referring to an oscillator, direct digital synthesis
dll	Dynamic link library – reusable software block - can be called from other programs
DR3	3rd order dynamic range receiver test - needs to be tested differently for an SDR
DSP	Digital signal processing
DUC	Digital up conversion - the transmitter part of a direct sampling SDR
DX	Long distance, or rare, or wanted by you, amateur radio station
FFT	Fast Fourier Transformation – convert time domain to frequency domain (and back)
FFTW	Fastest Fourier Transformation in the West – a software library used by many SDR programs
FIFO	First In First Out – a type of data buffer
FPGA	Field Programmable Gate Array – a chip that can be programmed to act like logic circuits, memory or a CPU.
Frame	Data packet structure – usually includes synchronisation, control and payload information in bytes
GUI	Graphical user interface – the form on the PC software that allows you to interact with the radio. The GUI is the knobs and display of an SDR.
HF	High Frequency (3 MHz -30 MHz)

Glossary of abbreviations and acronyms

Term	Description
I data or stream	The non-phase shifted part of the IQ audio signal or data stream
IF	Intermediate frequency = the Signal – LO (or Signal + LO) output of a mixer
Imaginary	Part of the complex signal in a buffer - often holds the Q signal or frequency bins below the LO frequency
IMD	Intermodulation distortion - interference / distortion caused by non-linear devices like mixers. There are IMD tests for receivers and transmitters. IMD performance of linear amplifiers can also be tested.
KK	Kiss Konsole – SDR application for Windows used as an example of PC software
ksps	Thousands of samples per second
L band	Amateur radio 1296 MHz band (1,240 to 1,300 MHz) also known as the 23cm band
LF	Low Frequency (30 kHz -3 MHz)
LO	Local Oscillator - the frequency that the SDR is tuned to. Usually the centre frequency of the spectrum display (panadapter)
MDS	Minimum discernible signal – measurement of receiver sensitivity
MOX	Microphone operated switch (in Kiss Konsole this is PTT from the PC software)
Msps	Millions of samples per second
NCO	Numerically controlled oscillator - the local oscillator in a DDS type SDR (CORDIC software code in the FPGA)
Net	An on air meeting of a group of amateur operators
NPR	Noise power ratio or 'white noise' receiver test
Panadapter	A spectrum display of a section of the HF spectrum (often called a receiver)
Payload	Data bytes that contain the digital representation of the signal rather than control or synchronisation information
PC	Personal computer - usually with a Windows, Macintosh or Linux operating system
PTT	Press to talk - the transmit button on a microphone - PTT signal sets the radio and software to transmit mode
Q data or stream	The 90 degree phase shifted part of the IQ audio signal or data stream
QSD	Quadrature sampling detector - SDR receiver front end (often a Tayloe detector)
QSE	Quadrature sampling exciter - SDR transmitter front end (often a Tayloe detector working in reverse)
QSO	Amateur radio conversation or contact
QSY	Request or decision to change to another frequency
Real	Part of the complex signal in a buffer - often holds the I signal or frequency bins above the LO frequency
RF	Radio frequency
RMDR	Reciprocal mixing dynamic range – receiver test.
SAM	Synchronous amplitude modulation

Glossary of abbreviations and acronyms

Term	Description
Sked	A pre-organised or scheduled appointment to communicate with another amateur radio operator
Slice	The part of the spectrum that is within the selected audio bandwidth i.e. the part you bandwidth that you can listen to.
Superhet	Superhetrodyne receiver – conventional (non SDR) receiver architecture
TR	Transmit / receive change over relay (or a signal that controls the TR relay)
USB	Universal Serial Bus – a computer port interface (not upper sideband in this book)
VHF	Very High Frequency (30 MHz -300 MHz)
VNA	vector network analyser - mode to use the SDR as an antenna or hardware filter analyser
Word	A digital word is one or more bytes used to carry a particular piece of information. For example the Q signal between the ANAN or Hermes radio and the PC software is a stream of 24 bit (3 byte) words. The microphone signal is 16 bit (2 byte) words.

Acknowledgements

As an owner of a Hermes based ANAN-100 transceiver, I offer a huge thank you to the dozens of contributors to the OpenHPSDR team, especially Phil Harman VK6APH, Kevin Wheatley M0KHZ, Kjell Karlsen LA2NI and Abhi Arunoday who developed Hermes. Thanks also to Graham Haddock KE9H for the Alex filters and to Abhi Arunoday and all the folks at Apache Labs, who turned the designs into off the shelf transceivers. It is great to see continuous innovation and the development of advanced new functions. Phil Covington who along with, Phil Harman, and Bill Tracey formed the OpenHPSDR group in 2005 has been a major force in SDR development, Greg Smith ZL3IX helped with the Mercury receiver.

Gerald Youngblood AC5OG the founder of FlexRadio brought the first commercial SDR transceivers to market, opening up the SDR experience to thousands of hams. He wrote the early code for PowerSDR and the excellent; 'A software-defined radio for the masses' series. FlexRadio Systems produces excellent radios and have expanded the boundaries of SDR. Their new transceiver range includes major advancements in performance, hardware, and software.

Also thanks to the software developers who develop the excellent SDR software which we can download and use for free. I certainly appreciate the thousands of hours of work that you have done to provide amateur radio enthusiasts with a completely new approach to operating ham radio.

And finally thanks to Tony Parks KB9YIG who produces the Softrock kits, which have opened up SDR to tens of thousands of experimenters, including me.

Summary

Well if you have managed to get this far you deserve a cup of tea and a chocolate biscuit (at least). It is not easy digesting large chunks of technical information. It is probably better to use some of the more technical sections as a technical reference. Anyway, I hope you enjoyed the book and it has made understanding software defined radio a little easier. I sure learned a lot about SDR in the process of, researching and writing it. Initially I was surprised that there is pretty much no information available online which gives an understanding of SDR PC software like Kiss Konsole, but having spent many hours pulling the C# code apart and then writing about it I can see why nobody else attempted it. Of course more fully featured PC applications would be even more difficult.

It is easy to forget the people who have given countless hours of their own time for free developing the excellent amateur radio SDR hardware and software that we tend to take for granted. Sure some of them have turned their hobby into a business, but without that SDR ham transceivers and receivers would never have become mainstream and would only be available as circuit board kits to a very few dedicated enthusiasts. SDR is challenging the boundaries of ham radio and it is beginning to offer performance that equals and in some cases exceeds the very best conventional radios. Competition is driving innovation and I hope we will see the release of both better SDRs and better conventional radios over the next few years.

Thanks for reading my book!

73 de Andrew ZL3DW.

THE END

Made in the USA
San Bernardino, CA
19 July 2014